G000022056

PRIMARY DIRECTIONS

TO WHICH IS ADDED A DISCUSSION OF
PROBLEMATIC POINTS AND A COM-
PLETE SET OF TABLES NECESSARY
FOR THE CALCULATION OF
ARCS OF DIRECTION

BY

SEPHARIAL

AUTHOR OF
"COSMIC SYMBOLISM," "A MANUAL OF ASTROLOGY,"
"A MANUAL OF OCCULTISM," "THE KABALA OF NUMBERS,"
"KABALISTIC ASTROLOGY," ETC.

Astrology Classics

Originally entitled, "Directional Astrology".

This is NOT *Sepharial's* "Primary Directions Made Easy", *a different book.*

Page 8 is taken from
Raphael's Astronomical Ephemeris of the Planets' Places for 1916.

Thanks to Kris Brandt Riske at the AFA for her kind assistance.

*For students outside the UK, the latitude of 51° 30' N,
used frequently in this book, is that of London.*

ISBN: 1 933303 17 4

Published 2006 by
Astrology Classics

The publication division of
The Astrology Center of America
207 Victory Lane, Bel Air MD 21014

On the net at www.**AstroAmerica.com**

PREFACE

A WORD by way of introduction to this work may be necessary, inasmuch as it deals with a technical subject, and the scope and purport of it cannot very readily be apprehended by the casual reader. It is essentially a book for the astrological student. To the astronomer it is particularly informing in that it brings out the more scientific part of the subject and shows the mathematical basis underlying the " lucky hits " to which many of our astrological exponents have undisputed claim.

The general scope of this work embraces all that is essential to the art of " directing " as practised by Claudius Ptolemy and Titus de Placidus, and more recently by Sir John Wharton, Mr John Gadbury, Commander Morrison, R.N., and Mr. A. J. Pearce, all of whom pursued the same general principles of astronomical directing, and differed considerably in their application of the celestial arcs to the measure of time. These points are reviewed and critically examined in the following pages.

An attempt having been made to bring the Arabian system of a day for a year into accord with the astronomical system of a degree for a year, some suggestions have here been made as to their rapprochement, the feeling being that, where credit is claimed for one system over another by exponents of either, the probability is that there is truth in both and hence there must be a co-ordinating factor. In the attempt to scientifically extend our horizon to include a pre-science of coming events, we have primarily to remember that there are many ways up a mountain, but there is only one top. A study of these various methods may lead to the conclusion that they are all leading in the same direction. It is as if one should say there are three hundred and sixty paces from end to end of the path, and another should say that there are three hundred and sixty-five. Both may be right according to their count and the measure of their tread, but the actual length of the path will remain the same whatever they make of it. This pathway is that which a man has to travel from his cradle to his grave ; and there is nothing that concerns a man so vitally as that he should know its trend and gradient, its pitfalls and rocky eminences, in advance of his going, so that experience may be laid by the heels and made to serve instead of to subjugate. And in the direst extreme of human experience we have to remember that " the wise man foreseeth the evil and obscureth himself, while the ignorant pass on and are hurt."

I have used a well-known and thoroughly authenticated horoscope for purposes of illustration, and anybody following the rules here given in relation to that horoscope will have no difficulty in following them out in respect to any other horoscope. Particular care has been taken to define the principle underlying each operation, and to give a clean-cut rule of procedure. Unlike most authors, who proceed by befogging the mind of the student with technicalities and afterwards explaining them by means of an appendix, I have devoted the first chapters of my work to technical definitions which are essential to the proper under-standing of the subject ; and until these are clearly apprehended and understood, the student need go no further.

To save further expense and trouble, my publishers have completed my work by the insertion of a complete set of tables, which include tables of Right Ascension and Declination for every degree of the zodiac, together with the ascensional difference due to the latitudes of London, Birmingham, and Liverpool under the present obliquity of the Ecliptic ; also tables of Sines and Tangents, and tables of Proportional Logarithms. These are all that are essential to the present treatise, and in themselves constitute a very valuable addition to the volume. It is, of course, presumed that the student of " Directional Astrology " [Primary Directions] will have mastered the preliminary task of setting a

4

PREFACE

horoscope for any given time and place with adequate precision, and hence that he is familiar with the use of an ephemeris. The present work is intended to replace and supersede Prognostic Astronomy, which is now out of print.

Beyond this I have nothing to say, save that I trust to have done my work efficiently and to have left no point on which a reader need question me. In such case the work may be regarded as complete, and so I hope it will be found.

SEPHARIAL.

CONTENTS

CHAP.		PAGE
1. Astronomical Definitions		9
2. Example Horoscope		16
3. Directions in Mundo		21
4. Directions in the Zodiac		26
5. Zodiacal and Mundane Parallels		31
6. Order of Directing		35
7. Effects of Directions		38
8. Planetary Indicators and the Measure of Time		44
9. Illustration		50
10. Ptolemy and Placidus		54
11. Directions Under Poles		60
12. The Part of Fortune		67
13. Lunar Parallax and Semi-Diameter		71
14. Lunar Equations		77
15. Cusp Distances		81
16. Suggested Method of True Directing		87
17. Conclusion		94
APPENDIX—Tables of Sines, Tangents, etc.		99
Tables of Right Ascension and Ascensional Difference		147
Tables of Proportional Logarithms		155

D	Neptune.		Herschel.		Saturn.		Jupiter.		Mars.		
M	Lat.	Dec.	Lat.	Dec.	Lat.	Dec.	Lat.	Dec.	Lat.	Declin.	
	° ′	° ′	° ′	° ′	° ′	° ′	° ′	° ′	° ′	° ′	° ′
1	0 S19	19N27	0 S39	17 S21	0 S29	22N18	1 S15	4 S13	3N31	14N50	14N53
3	0 19	19 28	0 39	17 19	0 29	22 19	1 15	4 6	3 36	14 56	14 59
5	0 19	19 29	0 39	17 17	0 29	22 20	1 14	3 58	3 41	15 2	15 5
7	0 19	19 29	0 39	17 15	0 29	22 22	1 14	3 50	3 45	15 9	15 13
9	0 18	19 30	0 39	17 13	0 28	22 23	1 13	3 42	3 50	15 18	15 23
11	0 18	19 31	0 39	17 11	0 28	22 24	1 13	3 34	3 54	15 28	15 33
13	0 18	19 32	0 39	17 9	0 28	22 25	1 13	3 26	3 59	15 38	15 44
15	0 18	19 32	0 39	17 8	0 28	22 26	1 12	3 17	4 3	15 50	15 56
17	0 18	19 33	0 39	17 6	0 28	22 27	1 12	3 8	4 7	16 2	16 9
19	0 18	19 34	0 39	17 4	0 27	22 28	1 12	3 0	4 11	16 16	16 23
21	0 18	19 35	0 39	17 2	0 27	22 29	1 11	2 50	4 15	16 30	16 37
23	0 18	19 35	0 39	17 0	0 27	22 30	1 11	2 41	4 18	16 45	16 53
25	0 18	19 36	0 39	16 58	0 27	22 31	1 11	2 32	4 21	17 0	17 8
27	0 18	19 37	0 39	16 56	0 26	22 32	1 10	2 22	4 24	17 16	17 25
29	0 18	19 38	0 39	16 53	0 26	22 33	1 10	2 13	4 27	17 33	17 41
31	0 18	19 38	0 39	16 51	0 26	22 34	1 10	2 3	4 29	17 49	

D	D	Sidereal	⊙	⊙	☽	☽	☽	MIDNIGHT.	
M	W	Time.	Long.	Dec.	Long.	Lat.	Dec.	☽ Long.	☽ Dec.
		H. M. S.	° ′ ″	° ′	° ′ ″	° ′	° ′	° ′ ″	° ′
1	S	18 39 16	9♈45 14	23 S 6	17 ♏54 18	5 S 9	22 S 7	25 ♏13 55	23 S56
2	☉	18 43 13	10 46 25	23 1	2 ♐38 55	4 46	25 22	10 ♐ 8 30	26 23
3	M	18 47 9	11 47 36	22 56	17 41 39	4 3	26 54	25 17 15	26 56
4	Tu	18 51 6	12 48 47	22 50	2 ♑54 1	3 2	26 27	10 ♑30 41	25 28
5	W	18 55 2	13 49 58	22 44	18 5 57	1 48	24 1	25 38 35	22 9
6	Th	18 58 59	14 51 9	22 37	3 ♒ 7 28	0 28	19 55	10 ♒31 36	17 24
7	F	19 2 56	15 52 20	22 30	17 50 12	0N53	14 39	25 2 37	11 44
8	S	19 6 52	16 53 30	22 23	2 ♓ 8 26	2 9	8 43	9 ♓ 7 24	5 38
9	☉	19 10 49	17 54 40	22 15	15 59 27	3 14	2 33	22 44 40	0N31
10	M	19 14 45	18 55 50	22 7	29 23 14	4 7	3N31	5 ♈55 29	6 27
11	Tu	19 18 42	19 56 59	21 58	12 ♈21 50	4 45	9 15	18 42 42	11 55
12	W	19 22 38	20 58 7	21 49	24 58 38	5 8	14 26	1 ♉10 8	16 47
13	Th	19 26 35	21 59 15	21 39	7 ♉17 45	5 16	18 56	13 22 2	20 51
14	F	19 30 31	23 0 22	21 29	19 23 31	5 10	22 33	25 22 43	24 0
15	S	19 34 28	24 1 28	21 19	1 ♊20 7	4 51	25 11	7 ♊16 13	26 5
16	☉	19 38 25	25 2 34	21 8	13 11 25	4 19	26 41	19 6 9	26 58
17	M	19 42 21	26 3 39	20 57	25 0 46	3 36	26 58	0 ♋55 38	26 38
18	Tu	19 46 18	27 4 44	20 45	6 ♋51 3	2 44	26 1	12 47 19	25 5
19	W	19 50 14	28 5 47	20 33	18 44 41	1 45	23 52	24 43 23	22 23
20	Th	19 54 11	29 6 51	20 21	0 ♌43 41	0 40	20 40	6 ♌45 45	18 42
21	F	19 58 7	0 ♒ 7 53	20 8	12 49 51	0 S27	16 33	18 56 8	14 12
22	S	20 2 4	1 8 55	19 55	25 4 51	1 33	11 42	1 ♍16 12	9 4
23	☉	20 6 0	2 9 56	19 41	7 ♍30 25	2 37	6 20	13 47 42	3 31
24	M	20 9 57	3 10 57	19 27	20 8 19	3 33	0 38	26 32 29	2 S16
25	Tu	20 13 54	4 11 57	19 13	3 ♎ 0 29	4 20	5 S11	9 ♎32 33	8 4
26	W	20 17 50	5 12 56	18 58	16 8 55	4 55	10 53	22 49 47	13 37
27	Th	20 21 47	6 13 55	18 43	29 35 21	5 14	16 14	6 ♏25 46	18 40
28	F	20 25 43	7 14 54	18 28	13 ♏21 5	5 16	20 53	20 21 19	22 50
29	S	20 29 40	8 15 51	18 13	27 26 22	5 0	24 28	4 ♐36 3	25 44
30	☉	20 33 36	9 16 48	17 57	11 ♐50 1	4 24	26 35	19 7 52	26 59
31	M	20 37 33	10 17 45	17 40	26 29 0	3 31	26 55	3 ♑52 45	26 22

Primary Directions

CHAPTER I

ASTRONOMICAL DEFINITIONS

THE following definitions must be fully understood by the student before the more intricate part of the system of directing is undertaken.

Longitude is of two kinds : longitude in the Orbit, and longitude in the Ecliptic. The latter is the only one recognised and used in this system. It is defined as distance from the vernal equinox, Aries 0, measured on the plane of the Ecliptic or Sun's path.

Latitude.—Celestial latitude is distance north or south of the Ecliptic.

Declination is distance north or south of the Equator. The Ecliptic lies in declination 23° 27' north and south.

Right Ascension is distance from the vernal equinox measured on the plane of the Equator. Right ascension thus answers to geographical longitude in the same way as declination answers to geographical latitude.

Meridian Distance is the distance of a celestial body from the midheaven of a place ; that is to say, from its meridian, measured in right ascension.

Semiarc of a planet is half the time it remains above or below the horizon of a place, measured in degrees of right ascension. The diurnal semi-arc is half the arc in right ascension of a planet above the horizon, and nocturnal semiarc is half the time it is (measured in right ascension) below the horizon. The diurnal semiarc taken from 180° will give the nocturnal semiarc, and the nocturnal semi-arc taken from 180° will give the diurnal semiarc.

Horizontal Arc is the distance in right ascension from a body to the point of its rising or setting. The semiarc less the meridian distance is always the horizontal arc.

Oblique Ascension is the right ascension of a body increased or diminished by its ascensional difference, according as its declination may be south or north. In northern latitudes the right ascension is increased for a body having south declination and decreased for a body having north declination, but the reverse of this is the case in southern latitudes.

Ascensional Difference is, the time (measured in right ascension) that a body is above or below the horizon more or less than six hours. If, therefore, its semiarc is more than 90° the excess of 90° is its ascensional difference. All bodies that are not exactly on the equinox (Aries 0 or Libra 0) have ascensional difference. For a planet in south declination the ascensional difference is added to its right ascension to get its oblique ascension, and for bodies having north declination the ascensional difference is subtracted. The reverse of this gives the oblique descension. The O.A. plus or minus 180° gives the obl. descension of the opposite point.

Pole of Latitude.—The pole of a place is the same as its latitude. The pole of a planet is measured by a circle of position or small circle parallel to the meridian of a place. The pole of the ascendant is the same as the latitude of the place, and this diminishes as we reach the meridian, where it is 0.

Direction is the process by which we bring the body of a planet to the longitude or body of another in a different part of the heavens either by its rising or setting, and this direction of one body to another, or to the place of another, is measured in right ascension ; that is to say, by the number of degrees which pass under the meridian of a place in the interval. All directions are taken in the prime vertical, or circle of observation—that in which a person stands upright facing south. Having the proportional distance of a planet between the meridian and horizon, we may bring another body to the same proportional distance along its own arc until it appears to be in the same relative position as the first body. This supposes that the position and influence of a planet is indelibly located in that part of the heavens in which it was found at the moment of birth. All arcs of direction are measured in right ascension.

Significators, in this scheme, are the Midheaven, Ascendant, Sun, and Moon. These are the bodies or positions that are directed or moved in the prime vertical in order to form conjunctions, oppositions, and various aspects with other positions and bodies. They are called " significators," from the fact that they are found to signify certain things in the life of an individual ; as, the Sun signifies male relationships, the Moon female relationships, the Midheaven honour and position, credit, etc., and the Ascendant the health and general play of events in the individual sphere of life. For further elaboration of this point refer to the *Textbook of Astrology* or *The New Manual of Astrology.*

Promittors.—These are the planets Neptune, Uranus, Saturn, Jupiter, Mars, Venus, and Mercury. The Sun and Moon may also be classed as promittors when the Midheaven or Ascendant is directed to them.

Logarithms, invented by Baron Napier of Merchiston, first-class mathematician and astrologer, were designed for the purpose of simplifying calculations in spherical trigonometry. In this scheme the arc of 90° of a right sphere is made to equal 10.00000, which is called the radix. Then, having the logarithm of any arc, it may be multiplied into any other arc by simple addition of their logarithms ; and, similarly, arcs may be divided by one another by subtracting one logarithm from another. Napier thus emphasises the fact that multiplication is merely the addition of a number to itself a given number of times, while division is merely subtraction a number of times. Then by means of a proportional circle we can multiply and divide any arc by simple addition and subtraction. The complement of an arc is what it lacks of 90°, and as this is equal to the radix 10, the complement of a logarithm is what it lacks of 10. Thus the logarithm of the sine of 32° is log. sine 9.72421, which is also the log. cosine of 58°, because 58 is the complement of 32, both together making 90. The arithmetical complement of the logarithm is 0.27579, since this, added to the log. sine of 32°, makes 10.00000. Familiarity with the use of logarithms will readily establish their great value in all mathematical calculations connected with the sphere.

I may now ask the reader to take in hand an ephemeris for the current year, 1916, and turn to the 1st January, and the above definitions may then be illustrated.

Let us suppose that a birth took place at noon, Greenwich mean time, on that date in London. The ephemeris being calculated for mean noon at Greenwich, there will be no equation of time necessary. The Sun, Moon, and planets will be in the positions indicated in the ephemeris. The Sun's longitude is seen to be Capricornus 9° 45' 14". The Sun never has latitude, inasmuch as it defines the Ecliptic, distance above or below which constitutes celestial latitude. All other bodies have latitude except when they are on that point where their orbits cross the Ecliptic, that is, their nodes. The course of the Sun being across the plane of the Equator at an angle of 23° 27' it will attain that declination at the solstices ; that is to say, on the 21st June and the 22nd December. On the 1st January it is found to have declination 23° 6' south of the Equator, and, therefore, would be immediately overhead at noon at a place which had geographical latitude 23° 6' south, and the Sun's diurnal course around the Earth would follow this parallel of latitude. The Sun's right ascension (R.A.) can be found in the tables (see Appendix) from its longitude.

Rule 1.—To find the R.A. of any body without latitude.

From the log. cosine of its distance from the nearest equinox subtract the log. cosine of its declination. Remainder is the log. cosine of its R.A. from the same equinox.

12

Example : The Sun is here 80° 15'
    from Aries 0   cos. 9.22878
    Its declination is 23° 6'   <u>cos. 9.96370</u>

Distance in R.A. from
    Aries 0 =79° 23'   cos. 9.26508

Therefore from 360° take 79° 23', and the R.A. of the Sun is thus found to be 280° 37'. Note that it is sufficient for our purpose to take the various quantities to the nearest minute of space.

Now take the Moon's place in the ephemeris, which is seen to be Scorpio 17° 54'. This is 47° 54' from Libra 0. The declination of the Moon is 22° 7'. Reference to the tables will show that the declination of Scorpio 17° 54' is 17° 10' only, and we therefore know that the Moon has latitude and is not on the Ecliptic at this time. The ephemeris shows it to have 5° 9' of south latitude. In finding its R.A., therefore, we have to take this latitude into account.

Rule 2.—To find the R.A. of a body having latitude.

Add the log. cos. of its distance from the equinox to the log. cos. of its latitude, and from the sum subtract the log. cos. of its declination. The remainder is log. cos. of its R.A. from the same equinox.

Example : Moon's distance from
        Libra 0=47° 54'   cos. 9.82635
        Its latitude is 5° 9'   <u>cos. 9.99824</u>

           Sum   cos. 9.82459

Moon's declination,   22° 7'   <u>cos. 9.96681</u>

Its R.A. from Libra   0= 43°53'   cos. 9.85778
R.A. Libra   <u>0=180°00'</u>
Moon's R.A.   =223°53'

Note.—If we take the arithmetical complement of the log. cos. of the declination and add it to the log. cos. of both the latitude and the longitudinal distance, we shall have the same result.

The R.A. of the other bodies is taken in the same manner, as they all happen to have some measure of latitude. Only when a body is in its node, and therefore coincident with the Ecliptic, does it have no latitude. In such case its R.A. is the same as that of the degree of the Ecliptic it holds.

We have next to find the meridian distances of the several bodies. To do this we have to find the R.A. of the Midheaven and Nadir, and take the nearest distance in R.A. of each body. Thus at noon on the 1st January 1916 the sidereal time is 18h. 39m. 16 secs. Convert this into degrees and minutes of the circle, thus : multiply the hours by 15 and call them degrees ; divide the minutes of time by 4 and call them degrees and minutes of space ; also divide the seconds of time by 4 and call them minutes and seconds of space.

Thus 18h. =		270°	0'	0"
39m.	=	9°	45'	0"
16s.	=	0°	4'	0"
R.A. of M.C. =		279°	49'	0"
		180°	0'	0"
R.A. of I.C. =		99°	49'	0"

The upper meridian is called the Midheaven (*medium coeli*) and the lower meridian is called the Nadir (*imum coeli*).

Having the R.A. of the M.C. and I.C., we are able to find the quantity of R.A. which separates the various planets from them, and this is the meridian distance of each of such planets.

Thus the Sun's R.A. was found to be 280° 37', and that of the M.C. (to which it is nearest) is 279° 49'. The difference is 0° 48', which is therefore the meridian distance of the Sun.

The Moon is found to be in the South-west quarter of the heavens, and therefore nearer to the upper than the lower meridian. Its meridian distance must therefore be taken from this point. Thus :

R.A. of M.C. =	279° 49'
R.A., Moon =	223° 53'
Meridian distance of Moon =	55° 56'

The other bodies are taken in the same way according to which meridian (upper or lower) they are nearest in R.A.

The semiarcs of the planets and luminaries have next to be found.

Rule 3.—To the log. tangent of the latitude of place for which the figure is set, or the horoscope cast, add the log. tangent of the planet's declination. The sum is the log. sine of the ascensional difference of that planet under the latitude of birth.

Uniformly, add this ascensional difference to 90° when the planet's R.A. is less than 180°, and subtract it from 90° if the planet's R.A. is more than 180°. The result is the diurnal semiarc of that planet. By subtracting this from 180° you will have the nocturnal semiarc.

Finally, by taking the meridian distance of the planet from its semiarc (diurnal if above the horizon, and nocturnal if below), you will have the horizontal arc, or distance in R.A. from the horizon.

Next find the proportional logarithm of the semiarc of each body, and take its arithmetical complement. Add to this A.C. the proportional logarithm of the planet's meridian distance. This is the constant log. of the planet for purposes of directing.

Enter all these elements into a single table, which is called the Speculum, an example of which will be found in the following pages. The scheme will now be ready for the practice of directing.

CHAPTER II

EXAMPLE HOROSCOPE

FOR the purpose of illustrating the method of directing by proportional semiarcs, I have selected the horoscope of John Ruskin, whose *Fors Clavigera, Mornings in Florence,* and other world-renowned works have stamped him indelibly as artist and man of letters as well as an independent thinker of considerable virility.

He was born at 7.30 in the morning of 8th February 1819, in London.

It is an invariable rule in practice to use that semiarc and meridian distance which are related to one another. Thus the Sun in the speculum is just below the east horizon at the moment of birth, as may be seen by comparing its nocturnal semiarc with its distance from the lower meridian, which are 110° 1' and 108° 44' respectively. This shows the Sun to be 1° 17' below the horizon. But as by the diurnal rotation of the earth on its axis from west to east the Sun will be carried above the east horizon upwards towards the Midheaven, during the course of which it will pass the places of Mars, Mercury, Venus, Neptune, and Uranus, it will be convenient also to have the semidiurnal arc and the meridian distance from the Midheaven. For whenever we use the nocturnal arc we always use the corresponding meridian distance from the lower meridian, and whenever we use the diurnal arc we also use the corresponding meridian distance from the Midheaven or upper meridian. This point should not be forgotten. It cannot be overlooked if the constant log. of the planet is inserted in the speculum, because this embodies the proportion of the semiarc to the corresponding meridian distance.

16

EXAMPLE HOROSCOPE

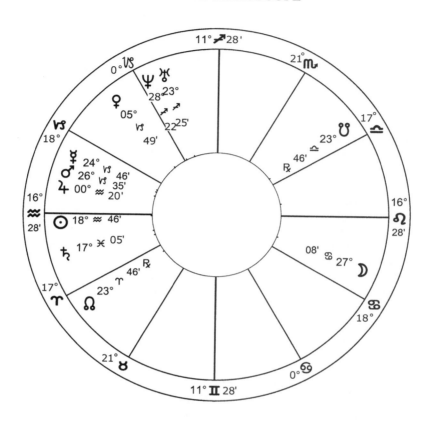

John Ruskin
8 February 1819   7:30 am   London

SPECULUM

Planets	Lat.	Declin.	R.A.	Merid. Dist.	Semi-arc.	Hor. Arc.
	° '	° '	° '	° '	° '	° '
Sun	—	15 13 S	321 12	108 44	110 01	1 17
Moon	5 01 N	25 39 N	120 17	50 21	52 51	2 30
Mercury	0 23 S	21 34 S	296 47	46 51	60 11	13 20
Venus	5 10 N	18 10 S	276 06	26 10	65 37	39 27
Mars	0 55 S	21 45 S	299 06	49 10	59 53	10 43
Jupiter	0 21 S	20 26 S	302 37	52 41	62 03	9 22
Saturn	1 56 S	06 54 S	348 54	81 02	98 45	17 43
Uranus	0 06 S	23 24 S	262 49	12 53	57 01	44 08
Neptune	1 13 N	22 14 S	267 47	17 51	59 05	41 14

17

Ruskin was agreeably disposed towards the idea of planetary influence in human life, for, although he confessed entire ignorance of the subject himself, he was always willing that others should have the full benefit of his experience and views, and he readily gave his birth data to those who sought it for the purpose of astrological calculations. His assertion that " there is more in it than is generally supposed " was doubtless the opinion he formed of the science from experience ; and if it does not carry the weight of scientific criticism, it stamps Ruskin, at all events, as a man of fearless integrity of thought.

In this horoscope we have a remarkable illustration of the principles of astrology. The Sun and Jupiter are rising in the humane sign Aquarius, while most of the planets are rising and above the horizon. These are indications of success and distinction in the world. The conjunction of Mars and Mercury in opposition to the Moon indicated that asperity and outspokenness which characterised this man of genius and rendered him fearless in the expression of his views. His eccentricities may well be attributed to the meridian position of Uranus and Neptune, while Venus, in closest aspect to Jupiter, and well elevated, disposed to success in the pursuit of art, of which he became a foremost exponent. But, of course, these positions do not make character. They only afford the opportunity for its full expression. Character and environment together constitute destiny, and it is undoubtedly often the case that one or the other of them is a misfit. It is only when we get a strong innate character with appropriate celestial environment that we look for the expression of genius.

We may now proceed to use this horoscope to illustrate the principles of directing.

Take first the Midheaven. This is directed by right ascension, and the planets coming to the meridian will form arcs of direction to it. The aspects to Midheaven should be noted. Thus the semisquare aspect falls in Capricorn 26° 28', and the sextile aspect is Aquarius 11° 28', and planets coming to these points will form aspects in the zodiac to the Midheaven. The square aspect falls in Pisces 11° 28' ; and

18

as Saturn is lower in the heavens than that point, it must come up to the place of this aspect and form the zodiacal square to the Midheaven. And the times in which these aspects are formed by the several planets will be in the proportion of their semiarcs. These directions are in zodiac.

The other kind of direction is in mundo—that is, in the circle of observation or prime vertical. Thus a body that is on the cusp of the twelfth house is in mundane sextile to the Midheaven or upper meridian, and one that is on the cusp of the eleventh house is in mundane sextile to the horizon or Ascendant. A planet that is in the middle of the eleventh house will be half way between the Midheaven and Ascendant, and, therefore, in semisquare aspect in mundo, because the meridian and horizon are always at right angles to one another. If a planet is not thus situated at the moment of birth it will afterwards attain that position, and the number of equatorial degrees which pass under the meridian from birth to the time when the aspect is formed will be the measure of the arc of direction. The original position of a body, either in the zodiac or in mundo, is always that to which direction is made.

Mundane Directions are those that are made to the apparent place of a celestial body, or to its aspects, in the circle of observation.

Zodiacal Directions are those which are made to the geocentric longitude of a body, or to aspects of that longitude, in the circle of observation or prime vertical.

All directions are formed by the rotation of the Earth upon its axis from west to east, by which the planets appear to rise, culminate, and set, pursuing a course that is from east to west. The lines or arcs traversed by the planets in this apparent motion are parallels of latitude of the same quantity and denomination as geographical parallels of latitude—that is, lines parallel to the Equator. The planets follow the parallel of declination in which they are found at the time of birth.

PRIMARY DIRECTIONS

It is understood that the radical imprint of a planet is localised in that part of the heavens it occupied at the moment of birth ; and although the actual planets do thereafter change their declinations and semiarcs, as well as their meridian distances, the radical imprint of the planet remains ever the same, and is to be regarded as entirely distinct from the planet itself, which, of course, moves along its arc in the heavens.

In the process of directing we are, therefore, only concerned with the radix or root horoscope and the changes which thereafter take place in the heavens, not among the bodies themselves, but in their relations to the radix. All directions of this nature are formed within a few hours of the moment of birth.

Directions (whether in the zodiac or mundo) are of two orders. These are " direct " and " converse."

Direct directions are such as are formed by one body being carried by the motion of the Earth towards another body or aspect in the heavens that precedes it. Converse directions, however, are such as are formed in the opposite direction. Thus in the foregoing horoscope of Ruskin, if we bring the Sun to the place of Jupiter, or Mars, or Mercury, or Venus, these would be direct directions, because that is the direct motion of the bodies in the heavens. But if we brought the Sun to the place of Saturn it would appear that we are carrying it backwards to a position that it held previous to the moment of birth. This, however, is not the case. The Sun is joined to the Earth by a direct ray which is called its earth-line, and it is this line which, by the rotation of the Earth on its axis, is carried down (bearing the solar imprint) to the place held by Saturn at the birth. This is a converse direction. But if we bring Saturn up to the place of the Sun it would be a direct direction.

Therefore all directions are formed by the one natural fact of the Earth's rotation on its axis, and aspects that cannot thus be formed are not within the category of primary directions.

We may now pass on to illustrate the method of forming every kind of direction, direct and converse, in zodiac and mundo.

CHAPTER III

DIRECTIONS IN MUNDO

THE principle involved in this process is that which enters into the construction of the horoscope, wherein we take one-third of the Sun's tropical semiarc as the extent of the house or division of the prime vertical. This principle enters into the construction of the tables of houses for various latitudes, the Sun's extreme declination remaining a constant quantity.

But in every horoscope we have the various planets with different declinations, and therefore with different semiarcs ; and consequently we are dealing with arcs which, although parallel to the Equator and to one another, are not parallel to the circle of observation. Hence an equal division of the prime vertical into twelve parts or houses will not effect an equal division of the various planetary semiarcs, which cut the meridian and horizon at varying angles depending on their declinations. Nevertheless, it has been found in practice that one-third of the semiarc, great or small, is equal to a house-space under the pole of that planet.

Suppose a planet to be exactly rising at the time of birth. Let its semidiurnal arc be 66° 21'. This is an arc of right ascension. Therefore when it has traversed one-third of its arc from the horizon to the meridian, 22° 7' will have passed under the meridian, and that will be the arc of the planet's direction to the cusp of the twelfth house. Another 22° 7' will bring it to the cusp of the eleventh house, and yet another arc of the same value will bring it to the meridian. When on the cusp of the twelfth house it will be in sextile aspect to the Midheaven, and when on the cusp of the eleventh it will be in sextile to the Ascendant, both directions being *in mundo*, as distinguished from similar aspects in the zodiac.

21

PRIMARY DIRECTIONS

If the Sun or Moon happen to be exactly on the cusp of a house, then the planet coming to the cusp by one-third divisions of its semiarc will simultaneously form aspects in mundo to the Sun or Moon. But if they are not so placed, then we have to find their proportional distances from the nearest cusp or limit of a house, and bring the planet to the same proportional distance in order to form the aspect.

Rule.—To find the cuspal distance of a planet. Note the cusp to which it is nearest at the time of birth. The distance of that cusp from the horizon compared with the planet's horizontal arc will give the planet's cuspal distance.

Example.—In the specimen figure the Sun is nearest to the cusp of the first house or ascendant, and therefore its horizontal arc, 1° 17', will be its cuspal distance. The Moon has a semiarc of 52° 51', and its horizontal arc is 2° 30', and as this is nearest to the cusp of the seventh house, that will also be its cuspal distance.

Now, as in all directions, the body to which direction is made is considered to remain stationary while the body directed is moved towards it by its natural motion in the heavens, we here direct the Moon to the sesquiquadrate aspect of the Sun, which it attains in the middle of the fifth house, that point being four and a half houses, or 135°, from the ascendant. The Sun, however, is not on the ascendant, and therefore we have to bring the Moon to a proportional distance from the middle of the fifth house. Thus

As the semiarc of the Sun, 110° 1',	
prop. log.	<u>0.21381</u>
arith. comp.	9.78619
Is to its cuspal distance, 1° 17',	2.14693
So is the semiarc of Moon, 52° 51',	<u>0.53223</u>
To its proportional distance, 0° 37',	
prop. log.	2.46535

Now, as one-third of the Moon's semiarc is 17° 37', that will be its house-space, and one-half will be 8° 48½', making for one and a half houses 26° 25½', and from this we subtract the above proportional distance, namely 0° 37', and there remains the arc of direction : Moon, 135°, Sun in mundo, 25° 48½',

Another example : Bring the Sun in the example horoscope to the mundane conjunction with Jupiter.

In order to effect this the Sun has to cross the horizon, its distance from which has been found to be 1° 17'. Thereafter we employ its diurnal arc and bring it to an equivalent distance from the horizon southwards as Jupiter is in the horoscope, by proportion of their semidiurnal arcs.

Jupiter's semiarc is 62° 3', and its meridian distance 52° 41', their difference 9° 22', which is the horizontal arc of Jupiter and therefore its distance from the cusp of the first house. Then we say :

As the semiarc Jupiter (arith. comp.) is to its cuspal distance, so is the semiarc of the Sun (diurnal =69° 59') to its proportional distance from the same cusp southwards. This works out as follows :

S.A. Jupiter, 62° 3'	log.	0.46253
Arith. comp.		9.53747
Cusp. distance, 9° 22'		1.28369
S.A. Sun, 69° 59'		0.41028

Sun's prop. distance =	10° 34'	log.	1.23144
Sun to horizon =	01° 17'		
Arc of direction =	11 51'		

Sun conj. Jupiter *m.*

It should be observed that the arc of direction to the horizon must always be added when the planet or body has to cross the horizon in forming the direction. Here the proportion of the Sun's arc to that of Jupiter gives a cuspal distance of 10° 34', and to this has to be added the distance of the Sun from below the horizon, making the arc altogether 11° 51'. When crossing the meridian to form a direction, no change of arc is necessary, but the arc to the meridian, which is the meridian distance of the planet, must be added to the arc formed on the other side of it.

It should be observed also that the body to which direction is made, and which is supposed to be stationary, supplies the first and second terms of the proportion, while that body which moves to form the direction supplies the third term and the resulting fourth term. In practice it will be found expedient to arrange all the mundane aspects in the order in which they are formed by each of the planets. The Midheaven and Ascendant remain stationary, and the Sun, Moon, and planets are the promittors that are moved to form directions upon them. Take one of these bodies at a time and make a list of the mundane directions it forms to the Midheaven, Ascendant, Sun, and Moon, calculate them, and arrange them afterwards in the order of their values. Always remember that the diurnal motion of the Earth upon its axis from west to east is the underlying cause of all directions, and that the planet to which direction is made, remains still, while the other moves towards it. You cannot then go wrong in your application of the method.

Direction to the conjunction in mundo is effected by bringing the body of a planet to the body of another, and not to its zodiacal longitude merely, as is done in the case of the zodiacal conjunction.

Thus in the case of Uranus to conjunction M.C. in mundo, we take its meridian distance as the arc of direction, whereas in the zodiac we take the meridian distance of its longitude, Sagittarius 23° 25', and this will be the arc of direction.

In all cases we bring the *body* of the planet directed to the conjunction or aspect of another body in mundo, to form mundane directions, all such directions being formed in the prime vertical, and expressed in terms of right ascension.

It will be found convenient to have the constant log. of the cuspal distance of each planet in the speculum. Subtract the proportional log. of the semiarc from the proportional log. of the cuspal distance. This will give the constant log., to which we have merely to add the proportional log. of the semiarc of any other planet to find the proportional cuspal distance of that planet.

It has been customary to regard the semiarc of a planet as equal to the quadrant, and therefore one-third as equal to a house or 30°. This is true in regard to a prime vertical whose pole is the same as the declination of the planet, but it is not true in regard to any other pole or geographical latitude. That is why we take the proportion of the semiarcs in finding the cuspal distances of planets. The test is this : If we take the oblique ascension of a planet, that is, exactly one-third of its semiarc from the horizon, it should have the same oblique ascension as the cusp of the twelfth house, but by adding 60 to the right ascension of the Midheaven to get the oblique ascension of that house, we shall find that if the planet has any other declination than 23° 27' there is a difference between the two results. It cannot, therefore, be truly said that a planet is in mundane sextile aspect to the Midheaven when it is one-third of its semiarc above the horizon, inasmuch as its position in the prime vertical does not then coincide with the cusp of the twelfth house ; but it may be said to correspond with that cusp on the general proposition that all circles are equal to one another and therefore that all quadrants are equal, and in practice it is found that one-third of a semiarc corresponds with one-third of the prime vertical, and this was allowed by Placidus, who was the first exponent of this system of mundane directions.

CHAPTER IV

DIRECTIONS IN THE ZODIAC

THESE are calculated on the same principle as mundane directions, that is to say, by proportion of the semiarcs; but instead of taking the actual body of the planet, or its position in the prime vertical, we take the longitude only and direct to that, and also to its aspects in the zodiac.

Thus in the horoscope of Ruskin the planet Neptune holds the longitude Sagittarius 28° 22', and therefore its zodiacal sextiles will fall in Aquarius 28° 22' and Libra 28° 22', its squares in Pisces 28° 22' and Virgo 28° 22', and so on.

The longitude of the planet, or its aspect if we are directing to it, remains stationary, and the actual body of the planet or luminary which is directed to it is moved along its own semiarc until it reaches the longitude or aspect to which direction is made.

Therefore we take the meridian distance and semi-arc of the ecliptic degree held by a planet and use these as the first and second terms of a proportion, in which the semiarc of the body directed forms the third term.

Example.—Direct the Sun to a conjunction with Jupiter in the zodiac.

Jupiter's longitude is Aquarius 0° 20', and from the tables we find this longitude to have R.A. 302° 31', from which take the R.A. of Midheaven, 249° 56', and we get its meridian distance, 52° 35'.

The same tables give the ascensional difference under London as 30° 51', which, taken from 90° as the declination is south, gives the diurnal semi-arc =59° 9'.

Prop. log. meridian dist.	52° 35' = .53442
" semiarc	59° 9' = .48332
Constant log. Aquarius	0° 20' = .05110
Prop. log. Sun's semiarc	69° 59' = .41028
" Sun's prop. dist.	62° 13' = .46138
Take from Sun's mend. dist.	71° 16'
Arc of direction	9° 3'

The constant logarithm of a longitude, once obtained, should be reserved, as it will serve for all zodiacal directions made to the same point of the ecliptic by simply adding the log. semiarc of the body directed to it. We then have the proportional meridian distance, which, compared with its original distance, gives the arc of direction.

Uniformly, find the R.A. of the longitude to which direction is made ; from this derive the meridian distance. Find its declination, and from this derive the semiarc. Subtract the proportional logarithm of the semiarc from that of the meridian distance, and derive the constant log. of the given longitude. To this constant log. add the log. semiarc of the body directed to it, and thus obtain the proportional distance of that body from the meridian at the point of direction. The difference between this and its radical meridian distance is the arc of direction.

Examples :—

1. Direct the Sun to aspects of the Midheaven in the zodiac. The aspects to which the Sun applies are the sextile in Aquarius 11° 28', the semisquare in Capricorn 26° 28', and the conjunction in Sagittarius 11° 28'.

Aquarius 11° 28' has R.A.   313° 55'
The Midheaven has R.A.   <u>249° 56'</u>

Merid: dist. of aspect   63° 59' prop. log. .44921
Asc. diff.   23° 09'
        <u>90° 00'</u>
Semiarc   66° 51'   prop. log. <u>.43017</u>

Constant log. of aspect in Aquarius 11° 28' =   .01904
Add prop. log. Sun's semiarc diurnal   <u>.41028</u>
Sun's prop. dist. from M.C.   66° 59'   .42932
Radical dist. of Sun from M.C.   <u>71° 16'</u>
Arc of direction, Sun sextile M.C. =   4° 17'

2. The next aspect of the Sun to the Midheaven in zodiac falls in Capricorn 26° 28', which is the semisquare aspect of 45°.

The R.A. of this longitude is 298° 29', and its meridian distance is therefore 298° 29' —249° 56' = 48° 33'. Its ascensional difference is 28° 40', which gives its diurnal semiarc =61° 20'.

Proportional log. 48° 33' —prop. log. 61° 20' =constant log. of aspect,   .10150
    To this we add the
    prop. log. of Sun
    as before, namely,   <u>.41028</u>

        .51178 =55° 23' Sun's propor. meridian distance ;
which take from   <u>71° 16'</u> Sun's radical distance,

remains   15° 53' the arc of direction Sun semisq. Midheaven.

3. The next aspect of the Sun to Midheaven in zodiac is the conjunction. For this the calculation is simply the difference of their right ascensions.

That of the Sun is	321° 12'
That of the M.C.	249° 56'
Difference	71° 16' = arc of direction.

These examples will doubtless serve for all cases that may arise in the course of directing a planet to the longitude and aspects of another in the zodiac.

We may now consider *converse* directions in the zodiac. These are calculated in exactly the same manner as the direct directions ; but instead of moving the directed body forward in the heavens, that is, from the Nadir to the Ascendant, from the Ascendant to the Midheaven, from the Midheaven to the Occident, and so on, we move it conversely against the natural diurnal motion of the celestial bodies in the heavens. Thus, in the example horoscope the Moon is in Cancer 27° 8'. Therefore, to bring Saturn to the square aspect of the Moon in the zodiac, we have to bring it to Aries 27° 8' by converse motion. We therefore find the meridian distance and semiarc of that point in the ecliptic, the meridian distance being taken from the lower meridian, to which it is nearest, and the semiarc being the nocturnal arc. Find the constant log. due to this point of the zodiac, and add to it the log. of the nocturnal semiarc of Saturn. From this we derive the proportional distance of Saturn from the lower meridian, and the difference between this and its radical distance is the arc of direction.

Similarly, we bring the Sun down the eastern heavens to form the converse zodiacal conjunction with Saturn. Here we take the meridian distance of Pisces 17° 5', and also its semiarc. Find the constant log. due to these and add to it the log. of the nocturnal semiarc of the Sun. The sum will be the prop. log. of the Sun's meridian distance at the conjunction, and the difference between this and the radical distance of the Sun from the same meridian will be the arc of direction.

PRIMARY DIRECTIONS

The bodies of Jupiter, Mars, Mercury, Venus, Neptune, and Uranus are brought to the zodiacal conjunction with the ascendant conversely by the measure of their horizontal arcs, which are derived by subtracting the meridian distance from the semiarc.

Thus Jupiter comes to the conjunction with the ascendant in zodiac conversely in an arc of 9° 21', Mars in an arc of 10° 43', Mercury in an arc of 13° 20', Venus in an arc of 39° 27', Neptune in an arc of 41° 14', and Uranus in an arc of 44° 8'. Similarly, the Moon is brought to an opposition of the ascendant in zodiac by an arc of 2° 30', which is the difference between its semiarc and meridian distance. This arc is much smaller than appears from its longitudinal position, and is due to the fact that the Moon has 5° of north latitude. A body with much north latitude sets much later and rises much sooner than does the degree of the ecliptic it holds. This is the radical difference between the mundane and zodiacal positions of a celestial body.

The Midheaven is directed to the conjunction with these planets in the zodiac by an arc equal to the difference of the R.A. of the Midheaven and that of the longitude of the planet.

Thus Venus comes to the Midheaven with the R.A. of Capricorn 5° 49', which is 276° 25', and the difference between this and the R.A. of the Midheaven 249° 56' =26° 29' arc of direction of Midheaven conjunction Venus in zodiac.

Uranus comes to the Midheaven in the zodiac by an arc of 12° 53', Neptune by an arc of 18° 17', Venus by an arc of 26° 29' (as above), Mercury by an arc of 46° 44', Mars by an arc of 48° 41', Jupiter by an arc of 52° 36', and the Sun by an arc of 71° 16'. These arcs, it will be observed, differ from the meridian distances of the several bodies as given in the speculum by an increment which is due to the latitudes of the various bodies. The meridian distances in the speculum will be the same as the measure of their directions to conjunction with the Midheaven in mundo.

We may now pass to another series of directions.

CHAPTER V

ZODIACAL AND MUNDANE PARALLELS

IN astrology the parallel of declination is deemed of the same significance and value as the conjunction, but its effects are more lasting, and if formed near the tropics, Cancer 0 or Capricorn 0, they will last for years together and characterise a whole period of the life.

A *zodiacal* parallel is formed by directing a body to the place held by a zodiacal degree which has the same declination as that held by a planet to which direction is made.

Example.—The Sun at birth has 15° 13' of declination. On the principle that all parallels of declination, being at the same distance from the Equator, act magnetically in unison, any body coming to an ecliptic degree which holds the same declination as the Sun, namely, 15° 13', whether north or south of the Equator, will act as if in conjunction with the Sun. Reference to the tables will show that there are four points which have this same declination, namely, Aquarius 18° 46', Taurus 11° 15', and Scorpio 11° 15'. Therefore, if we direct any body to any of these four longitudes in the zodiac by the rules given for directions in the zodiac, we shall bring them to parallels of the Sun in zodiac. The process is exactly the same as if we were directing to an aspect in the zodiac.

A *mundane* parallel is formed by the direction of a body to the same distance on one side of the meridian or horizon as that radically held by another body on the other side of the same meridian or horizon. These can be readily computed by reference to their horizontal arcs.

Example.—Bring Saturn to the mundane parallel of the Sun. The Sun's radical distance from the horizon north-wards is determined by the difference of its meridian distance and semiarc, namely, 110° 1' —108° 44'=1° 17', and we therefore have to bring Saturn to the same distance above the horizon. The semiarc of Saturn is 98° 45', and its meridian distance 81° 2' ; its horizontal arc therefore is 17° 43'. Then say : As the semiarc Sun is to its horizontal distance, so is the semiarc Saturn to its proportional distance, which, being added to the first or radical distance of Saturn from the horizon, will give the arc of direction.

Some writers on this subject have repudiated the parallel in mundo formed upon the horizon, but without adequate reason being adduced in support of their objection. Yet the same writers have not denied the efficacy of parallels formed on the *same* side of the meridian, one south and the other north, as in the 4th and 9th houses, or the 11th and 2nd, 10th and 3rd, etc., forgetting that bodies so placed are at equivalent distances from the *horizon* !

The rule for parallels is the same as for aspects. As the semiarc of the stationary body is to its meridian distance, so is the semiarc of the moving body to its proportional distance, which, taken from its primary distance, or added if it passes into another quadrant in forming the aspect, will give the arc of direction.

Thus we may bring Saturn to a parallel with the Moon in mundo. The Moon here is 2° 30' from the west horizon, and below it. If we bring Saturn along its own arc until it reaches a proportionate distance below the east horizon, we shall have a mundane parallel formed on the same side of the horizon, but on opposite sides of the meridian. We could work this problem by reference to the meridian distances of the two bodies from the Nadir, and the result would be the same.

It should be observed that the Sun and Moon are regarded as significators in the formation of mundane parallels by the other bodies, and the meridian and horizon therefore become sectors, upon which the parallels are formed.

ZODIACAL AND MUNDANE PARALLELS

Another form of the parallel in mundo is what is known as the rapt parallel. This is formed by the motion of the Earth on its axis, whereby the various bodies are carried from east to west at their several relative distances from one another until they come to the same distance on either side of the meridian or horizon. In this case *both* bodies move in the prime vertical at a rate proportionate to their relative semiarcs.

Rule.—As half the sum of their semiarcs is to half the sum of their meridian or horizontal distances, so is the semiarc of the body applying to the angle, to its distance from that angle at the formation of the parallel. This distance taken from its radical distance from the same meridian or horizon will give the arc of direction.

What we are actually doing is to bring the meridian or horizon to the mid-distance between the Sun and a planet, or between the Moon and a planet. And these mid-distances are of the greatest significance, whether in the zodiac or in mundo. Here we are considering them only in mundo.

Example.—Bring the Moon and Saturn to a rapt parallel. This is formed on the lower meridian.

Semiarc, Moon (nocturnal)	52° 51'		
" Saturn "	98° 45'		
	2)151° 36' *(division by 2)*		
Half sum of semiarcs	75° 48' prop. log.		.37560
		Arith. comp.	9.62439
Merid. dist. of Moon	50° 21'		
" " Saturn	81° 02'		
	2)131° 23'		
	65° 41'	prop. log.	.43782
Semiarc, Moon	52° 51'	"	.53223
Proportional dist., Moon	45° 48'		.59444
Radical distance	50° 21'		
Moon rapt. par., Saturn =	4° 33' arc of direction.		

PRIMARY DIRECTIONS

Note.—In all cases where the Midheaven (meridian) and Ascendant (horizon) are employed as sectors, the Sun and Moon are employed as significators. They form aspects by their own apparent motions in the prime vertical, and the planets form aspects to the radical of the Sun and Moon by the same motion. This is the underlying principle of all parallels in mundo, and all rapt parallels. Remember that in mundane directions you are always dealing with the bodies themselves and not their longitudes.

CHAPTER VI

ORDER OF DIRECTING

THE student will do well to employ some definite method of noting the various directions, and of collating and tabulating his results, otherwise he is sure to overlook some that are important when considered in association with others that attend them, whether they be of the same or a contrary nature. Thus, if in a train or sequence of evil directions there should occur a good aspect of Jupiter to the Sun or Moon, the health and fortunes will be greatly sustained thereby, so that what would otherwise appear a fatal set of arcs, in the presence of this benefit arc of direction would lose that extreme significance, and, although sickness might supervene, the good direction would indicate a favourable crisis.

The following method is therefore suggested as inclusive of all legitimate directions.

1. *Mundane Directions*

(a) Direct all the bodies to aspects and conjunctions with the Ascendant from east to west and from west to east.

(b) Direct each of the bodies to all the aspects and the conjunction with the Midheaven, both ways.

(c) Direct the Sun to other bodies and their aspects in mundo, both ways.

(d) Direct the Moon to other bodies and their aspects in mundo, both ways.

(e) Direct each of the planets separately to mundane aspects and conjunctions with the Sun.

(f) Do the same in regard to the Moon.

(g) Direct the Sun to mundane parallels with the Moon and planets.

(h) Direct the Moon to mundane parallels with the Sun and planets.

(i) Direct the Sun to rapt parallels with the Moon and planets.

(j) Direct the Moon to rapt parallels with the Sun and planets.

2. Zodiacal Directions

Follow the same order as for mundane directions, omitting classes (g), (h), (i), and (j) (mundane and rapt parallels), which are not formed in the zodiac.

Note that in zodiacal directions a body is always moved to a longitude to form a conjunction or aspect, never the reverse of this. Also that the meridian and horizon are fixed circles which do not move in regard to any particular locality. The Midheaven and Ascendant are those points where the ecliptic cuts through the meridian and horizon respectively.

All this long process of directing may appear to be very tedious. It certainly requires patience and method. But once done it lasts for a lifetime, which is a point to be considered. In possession of such a chart one may direct one's course with wisdom and success, avoiding those dangerous shoals, sandbanks, and breakers which occur in the course of every life—or, if it be beyond the power of a man so to do, he can at all events divest evils of much of their power over him by adjusting himself to them, making provision against times of evil fortune and doubling his efforts when times of prosperity are shown. Thus may a man order his going and bring his life to a peaceful end. Sudden death cannot overtake the man who has knowledge of the time of that event years in advance ; and the keen edge of many afflictions, to which an all-wise Providence may dispose us for the greater ends of life, are dulled by a philosophic anticipation, so that, cutting less deeply, they leave the vital soul of man unhurt. Therefore, rather than pray that what is foreordained by the laws of life to the inscrutable ends thereof may be averted, let us rather pray with the Psalmist : " Teach me the number of my days, that I may apply my heart to wisdom."

CHAPTER VII

EFFECTS OF DIRECTIONS

In order to complete this section of the work, which deals with that system of direction by semiarcs currently practised and approved, it will be necessary here to indicate the general effects of directions, so that the nature and import of events may be known as certainly as the time at which they are likely to transpire. I am here speaking of " effects " of directions as if these latter had a direct dynamic result upon the character and actions of an individual. I am disposed to classify astrologers in three main groups—fatalists, casuists, and idealists—according to the various views they take of the nature and purport of astrology. The Fatalists believe, or profess to believe, that there is a planetary configuration and an event which attends it. They admit no possible intervention, amelioration, or extenuation. *Che sarà sarà*, and that is the end of the matter. They argue a certain necessity of connection between character and environment as we find it and planetary positions at the moment of birth. As regards " directions," all of which are formed within a few hours of the birth, they speak of them as " seeds sown " in the plastic soil of the human soul which spring up and bear fruit at the appointed time, as measured by the arc of direction. They are born when they are born by necessity of universal law, and they die when they die because fatal arcs of direction are then in force.

They speak of laws of Nature as if they were dynamic forces against which mankind cannot possibly contend. They forget that laws are only mental concepts induced upon our minds by an apprehension of the correlated successiveness of events, and that what we know about natural laws is an infinitesimal part of the possibly knowable. They speak of the bodies of this microscopic solar system of ours as if they were the be-all and end-all of existence. They forget that the continuity of matter is a fact only on the material plane, and that there are forces of an immaterial nature which transcend both matter and what we call the laws of material existence. The moral law is an illustration of this. It is spiritual in its origin and spiritual in its effects. If astrology teaches fatalism, its use is at an end and it becomes a suicidal science, since there is no object in knowing that which must inevitably take place. It would reduce man to an automaton and divest him of all moral responsibility.

The Casuists are those astrologers who accommodate their facts and figures to popular concepts by a discreet use of a *mélange* of spurious philosophy. They forever quote the effete adage : " The wise man rules his stars, the fool obeys them " ; and that other which says : " The stars incline but do not compel." They put a premium upon the wisdom of experience and the will-power of a purposeful character, and promptly consign a man to destruction by telling him that his horoscope indicates he has neither one nor the other. They do not suggest to him that astrology, properly conceived and applied, is in itself the very concrete of experience, nor that the will-to-be and the will-to-do are functions of the human soul which rise superior to all circumstance, outlasting life itself.

The Idealists are those among astrologers who regard the intelligible universe as the expression of a Supreme Intelligence, who regard the planetary combinations merely as symbols, knowing that the causes of all effects are within man himself, the cogniser of all experience. They regard the " signs of the times " as the driver of a locomotive regards the signals, not as " causes " of disaster, but as warnings against it, an open book to those who can read the signals, but of no value to those who cannot. They look upon the science of astrology as a wireless operator looks upon his code-book, merely as a means of interpreting the signals—a science evolved by man for the service of man.

My own view of the matter is that there is some thing to say for the materialist side of the question, and a great deal more for the idealistic. There is not the shadow of doubt in my own mind as to the material fact of the interaction of the planetary bodies, nor as to the fact that this interaction is registered by an intervening body of the system only at certain angles. The Platonic dictum that " God geometrises " is nowhere better illustrated than in the law governing the interaction of bodies belonging to the same system. The physical effects of the syzygies, and especially of ecliptic conjunctions of the luminaries, are immediately appreciable. The law of the tides is a concrete example of the fact of interplanetary action. We cannot deny the dynamic effects of planetary action on the material plane, and we have every reason for including in this category the human organism, compounded as it is of cosmic elements and in direct physical relations with a material environment. But that does not warrant us in extending our views to include the action of physical bodies upon the immaterial part of us, the only part of us that is essential and distinctively human. The only thing that can directly affect the soul of man is the soul of another human being. There is continuity of action upon all planes of existence because there is a continuity of matter upon all planes, but we have no grounds for extending the range of action from one plane to another plane, except it be by mediation or agency. Else we could say that a good soul must be possessed of a sound body, a beautiful soul of a comely body, and that our moral principles are derived from what we eat and drink—instead of which, what we eat and drink depends on our moral principles. There is sound philosophy in the words of Tennyson when he says that " Soul to soul strikes through a finer element of its own." It is capable of acting mediately through the physical body or immediately through its own essential being. These views will doubtless alter our viewpoint in regard to much that hitherto has been regarded as fundamental to a belief in astrology. The effort to accommodate the facts of astrology to the materialistic science of a generation agone has tended to this issue. Without in any way disposing of astrology as a physical science, it is high time that we learned to interpret the facts of that science in the light of the higher spiritual teaching to which we have access. Otherwise we shall debase the science and enslave our own souls. In such case it were better that our astrology had never been written. As a physical science, astrology has an immense future before it in this utilitarian age upon which we have embarked ; but as a fatalistic creed it is not worth an hour's study.

These remarks will enable the reader to understand why, in the following statement of the " Effects of Directions," I have pursued the common practice of attributing certain results or sets of conditions as accompanying the formation of " directions " or planetary combinations in the horoscope subsequent to the birth. They should not be regarded as inevitable " effects " of such directions, but rather as things signalled, as if we should hoist the red light to indicate " danger ahead," the green light for " caution," and the white light for " road clear." These signals do not cause disasters, but our ignorance of them, our inability to see them, or our wilful disregard of them may very well result in a catastrophe. Human science has harnessed many of the subtle and intangible forces of Nature and deployed them to the service of man. It may do the same with cosmic forces that are as universal as etheric action.

The Midheaven

This point of the horoscope stands for dignity, influence, authority, and position, the worldly honour and credit of the subject, and for all that is associated with his social and communal status. Good directions, such as the sextile and trine of all planets, and the conjunction and parallel of Jupiter, Venus (and Mercury when well aspected at birth), are indications of an enhanced position, higher honours, social distinctions, increase of prestige, etc.

Evil directions, such as the semisquare, square, and opposition of all planets (including the Sun and Moon in this category), and the conjunctions and parallels of Uranus, Neptune, Saturn, and Mars, indicate assaults upon the good name and credit of the subject, hurt to the business affairs, loss of position, rivalries, and unprofitable associations.

PRIMARY DIRECTIONS
The Ascendant

This point of the horoscope indicates things personal to the subject, as health, general welfare, comfort, environment, changes, and the common relationships of life, that which affects him through collective influence, the public state of affairs, etc.

Good aspects (as above enumerated) tend to benefit the subject by a variety of means differing as the nature of the planet which is in aspect by direction.

Evil aspects signal bad health, obstacles, hindrances, incommodities, troubles and annoyances of various kinds, according to the nature and position of the planet directed.

The Sun,

when in a hylegliacal place (as defined by Ptolemy), has significance of the vital constitution and life of the subject. Generally it stands for the father and male representatives of a family, and for the honour, credit, and position of the subject himself. It is thus associated more particularly with the Midheaven.

The Moon

denotes the health, changes of fortune, the mother and female representatives of the family, the functional powers of the body, and, in its association with the Ascendant, public bodies, the populace, and public concerns generally.

If in a hylegliacal position, it indicates the vital organs and life of the subject.

Note.—Ptolemy defines certain parts of the horoscope as being vested with a vital prerogative, wherein the Sun has precedence by day and the Moon by night. It is a moot point whether other bodies, being in such positions (in the absence of the luminaries), may not be vested with the same prerogative, and again, whether the Sun or Moon, not radically in such a position, may become invested with such significance by coming to a hylegliacal place by direction after birth. Failing either the Sun or Moon, Ptolemy invests the Ascendant with the properties of hyleg or life-giver. But, whatever may be concluded in this debatable matter, it is certain that the Ascendant is most generally affected by evil directions at the time of a physical crisis, the afflicting planet generally indicating the nature of its cause.

The above points in the horoscope, the Mid-heaven, Ascendant, Sun, and Moon, are the significators, because they signify such persons and things in the life of the subject as are capable of being affected by the conflict of human circumstance.

All directions are made either (a) by the natural motion of the significators to the places and aspects of the planets, or (b) by the natural motions of the planets to the places and aspects of the significators.

The triangle (trine) and parts of it are good aspects, and indicate some advantage according to the position and nature of the planet directed. The cross (square) and parts of the square are evil aspects, and indicate similar disadvantages.

CHAPTER VIII

PLANETARY INDICATORS AND THE MEASURE
OF TIME

THE following definitions of planetary indications are necessarily only partial and incomplete, but they will serve doubtless to convey a more or less definite idea of the nature of events which may be expected to attend directions formed by them with the various significators.

It should be observed that the house which a planet directed to holds in the horoscope of birth, or that which a planet which is directed arrives at when the aspect is complete, has chief significance in regard to the department of life in which the events will transpire, the nature of those events depending primarily on (a) the nature of the aspect and (b) the nature of the planet involved.

In this light, therefore, it may be said that *Neptune* in good aspect indicates events of a beneficial nature connected with the use of the faculties or some special faculty, and frequently in connection with a form of art ; benefits from unexpected sources coming mysteriously to the subject ; unseen and intangible influences at work for the benefit of the subject ; brilliant flashes and inspirations of the mind ; spiritual aid ; intuitive activity.

In evil aspect by direction it denotes chaotic and mysterious events adverse to the interests ; scandal, secret enmity ; undermining of the credit by misrepresentation and fraud ; treachery, ambush ; an involved state of affairs ; nervous leakage and depletion of energy ; wasting of tissue ; physical ennui and decline of the vital powers from inscrutable causes ; apprehension, fear, and dread of consequence ; danger of espionage ; loss by fraudulent concerns and false investments ; mental unrest and loss of faculty.

Uranus in good aspect denotes civic and governmental honours, preference, advancement ; unexpected benefits arising out of public concerns and affairs ; ingenuity, inventiveness ; originality ; success in mechanical and engineering business ; strokes of good fortune coming from unexpected sources ; new associations and alliances.

In evil aspect this planet denotes the breaking down of existing relationships, lesions and fractures, partings and separations, loss of a sudden and unlooked-for nature ; hurt by strikes and public demonstrations ; nervous lesion, paralysis ; breaks and dislocations.

Saturn in good aspect indicates favours from aged persons and benefits from old associations, long investments, time contracts, and a general state of stability and steadiness in the fortunes, congenial retirement and sequestration.

In evil aspect Saturn depletes the vital powers, causes physical hurts by falls and contusions, morbid diseases, colds and chills, inhibition of bodily functions ; loss of money and property ; mental and nervous depression ; privations, obstructions, hindrances, and general misfortunes. Saturn is anciently known as the Greater Infortune.

Jupiter in good aspect denotes increase of fortune, opening up of new and lucrative opportunities, expansion of interests, advancement, progress, honours, confidence, good judgment, a general feeling of expansion and well-being, both physical and mental.

In evil aspect Jupiter denotes losses, errors of judgment, vanity or excessive confidence, disfavour of legal men and clericals, physical disabilities arising from congestion and surfeit, excess or over-indulgence, " too much of a good thing," too much *confiance en soi,* and consequent loss of esteem with others. It indicates a period of low finance, due to lavish expenditure, severe losses, or heavy investments. Jupiter is anciently known as the Greater Benefic, but it is certain that its evil aspects denote anything but a beneficial state of affairs.

Mars in good aspect denotes activity, new enterprises, great output of energy with commensurate good results, travelling, the executive powers are stimulated, and much profitable work is done.

Benefits accrue from military men, business connected with iron, steel, and fire. The muscular system is strengthened and there is a disposition to increased activity. Honours due to deeds of daring and chivalry. Women frequently marry under this aspect.

In evil aspect Mars denotes hurts by burns, scalds, fire, and steel, with loss of blood, abrasions and cuts, and also fevers and inflammatory conditions of the body or that part of it indicated by the position of Mars by direction. Loss by fire or theft, sometimes attended by violence. Sudden alarms and disasters of various sorts. Mars was anciently known as the Lesser Infortune.

The *Sun* in good aspect indicates increase of prestige, honours and emoluments, new friends and associations of a creditable character, general advancement and good fortune.

In evil aspect the Sun denotes losses, disfavour of superiors, troubles through male members of the family, the chief, overseer, or manager of a business ; loss by governing bodies ; ill-health due to fevers. Reverses of various sorts according to the house in which the direction is completed.

Venus in good aspect signifies social and domestic success, pleasures and enjoyments, gifts and presents, decorations ; the young court or marry, and the mature have children born or daughters engaged or given in marriage, and such events happen as cause pleasure and satisfaction. The affectional nature is stimulated and the health is good.

In evil aspect Venus denotes sorrows, disappointments, bereavements, grief, and losses, domestic and social troubles, and hurts associated with young women or children. Venus was anciently known as the Lesser Benefic, and the less one has of it when in evil aspect the better for all concerned.

Mercury acts in terms of the planet to which at birth it is in closest aspect ; but if not within orbs of an aspect with any planet, then in terms of the ruler of the sign it occupies. In good aspect it usually signifies activity, much occupation of a profitable nature, connected with writings, science, and business of a general nature. Travelling, profitable journeys, good news, gain in connection with the avocation or trade. An active time generally.

In evil aspect Mercury produces annoyances and disturbances, evil news, worry and anxiety, many short journeys to and fro to no purpose or profit, sleeplessness, irregular feeding, unrest.

The *Moon* in good aspect denotes pleasant and profitable changes, a change for the better in the general state of affairs, gain by public associations and concerns, favours from women of mature age, popularity.

In evil aspect it denotes loss by any of the above means, and a state of unrest both physical and mental which leads to neglect of duties and consequent loss. Hurts from women. Some public affronts may be suffered. Changes are unfortunate, and best avoided.

The Measure of Time

In the foregoing system of primary directions by proportion of the semiarcs, the measure of time is $1° = 1$ year, and every $5' = 1$ month.

Considerable discussion has been devoted to the question of time measurement in directions. Those who advocate the Arabian system of a day for a year have sought to bring primary directions into line with that system by equating arcs of direction made on the foregoing principle of semiarcs, by adding the arc to the Sun's right ascension at birth, and then finding the day after birth at which the Sun attains this new right ascension. The count is made at the rate of one day for a year of life, and two hours for every month.

Others, again, have sought to apply a plus increment at the ratio of 365 to 360, seeing that the Sun moves through 360 degrees in 365 days, which is the same as taking the Sun's mean motion 59' 8" as the value for 1 day = 1 year.

But it is obvious that none of these methods has any direct application to the system we are now concerned with, inasmuch as all the directions formed by the diurnal rotation of the Earth on its axis are formed within a few hours of birth so far as they apply to a life of ordinary length, and they are measured in degrees of right ascension—that is to say, by the passage of the Equator under the local meridian in the prime vertical,—and therefore degrees of right ascension are the only uniform basis of measurement. It certainly does not seem consistent to measure arcs by one method and equate them in terms of another.

It should be observed, however, that primary directions in right ascension do not always coincide exactly with the events they are held to signify. Sometimes they are too short, and sometimes too long, but never more than a few minutes either way. Commander Morrison, R.N., was of opinion that the event signified was delayed or accelerated by reason of current transits in the horoscope at the time, and he further states that the chief effects may be expected to transpire when the lunar or secondary directions come into accord with them. This gives rather a wide margin of operation to the primary direction, and has led many to the conclusion that the secondary direction is, after all, the important one and deserving of primary consideration. A very little experience will show that it is not so, for, unless there are concurrent primaries in operation, lunar or secondary directions frequently pass with little or no result.

Primary directions and transits appear to answer to all the more important events in life.

At the same time we have to consider the *duration* of effects, and in regard to this it has been observed that the process of formation of an arc of direction should be considered. For the longer a direction may be in forming, the longer will those events endure which it signifies. Here we have

Fitzroy's old maxim again in evidence : " Long foretold, long last : short notice, soon past."

Thus an aspect to the Ascendant formed during the rising of a sign of short ascension such as Aquarius, Pisces, Aries, Taurus in northern latitudes, and the opposite signs to these in southern latitudes, will be speedily formed and over. On the other hand, a similar direction formed to the Ascendant when in a sign of long ascension will be formed more slowly, and will dissolve more slowly. In such case we might expect the signified event to begin to happen earlier and to end later than in the former case.

One finds in experience, however, that men frequently trace years of toil and suffering to a sudden disaster over-taking them in a moment. In my theory of transits, this could not happen in earlier years, but might easily occur at maturity when the accumulated results of a man's labour were heaped around him. (See Transits.)

The following illustrations will, however, sufficiently prove that there is adequate coincidence between arcs of direction and the events they are held to signify, to warrant the measure of time $1^0=1$ year as scientifically valid.

CHAPTER IX

ILLUSTRATION

IN the example horoscope given in these pages we have a singularly interesting subject. The chief events of the life are well defined and closely indicated by the attendant arcs of direction. Hundreds of horoscopes, whether pertaining to individuals in high walks of life or of modest position in the world, could be adduced to show that this coincidence of direction and event is not fortuitous, but regular and consistent, and as dependable as any astronomical formulary. The student will find pleasure and instruction in working out the following arcs of direction in the present instance.

John Ruskin leapt into fame and became a " lion " in the world of art in the autumn of 1843 under the direction of

Sun sextile Midheaven mundo 24° 37'

He was married on the 19th April 1848, and, while on his honeymoon, took a chill while sketching in Salisbury Cathedral and was seriously ill. This happened under the adverse directions—

Moon oppos. Venus mundo conv.	29° 16'
Moon square Venus mundo dir.	29° 16'

The nearness of these adverse arcs to the event of marriage proved unfortunate for such a domestic change. The danger of his choice of a wrong time and a wrong partner for marriage was radically indicated by the Moon's opposition to Mars and Mercury, and nothing but constant bickering could have been expected from such indications.

The first serious break in Ruskin's health was in May 1840, for which we have the direction—

Moon oppos. Saturn zod.   21° 46'

He received honours from the University of Cambridge in May 1867 under the appropriate directions of

Ascendant trine Venus mundo 48° 2'
Ascendant sextile Moon zod.   48° 22'

He was elected Slade Professor of Fine Art on the 10th August 1869, and commenced his course of lectures under the following directions :

Ascendant trine Jupiter zod.   50° 35'
Midheaven par. Jupiter zod.   50° 57'
Ascendant par. Jupiter zod.   51° 14'
Moon rapt par. Jupiter   51° 22'

In the following year his health gave way, and his mother died in December of that year, 1871. The arc for that year measured from 51° 53' to 52° 53', and within these limits we have the significant directions—

Ascendant square Saturn zod. con.   51° 59'
Moon square Mars mundo   52° 0'
Sun par. Uranus zod. con.   52° 0'
Ascendant square Jupiter mundo   52° 41'

followed by Moon par. Mars zod. 53° 3', close upon the death of his mother.

His health completely broke down again in 1888, under the directions—

Sun opposition Uranus zod.   68° 49'
Sun opposition Uranus mundo   69° 14'
Moon rapt par. Saturn   69° 30'

Here the Sun is hylegliacal, and, being so heavily afflicted from angles of the horoscope, and the Moon also afflicted by Saturn, only disastrous illness and misfortune could have been signified.

Nevertheless, he survived this affliction, and further added to his reputation as a man of letters and exponent of fine art during some ten years, until eventually, with declining vitality laying him open to attack, he was afflicted by influenza and succumbed on 20th January 1900, the arc for that date being 80° 57'. The following significant train of directions was then in force :

Sun par. Uranus zod.	80° 10'
Ascendant square Saturn mundo	81° 2'
Ascendant sesquiq. Sun zod.	81° 11'
Ascendant par. Uranus zod.	81° 27'
Moon square Mars zod. con.	81° 30'

In view of these directions, it cannot be said that we are not duly signalled by the celestial bodies, not only of the approach of evil times, when more than usual care and attention are due to health and fortunes, but also of those periods of good fortune when the sun smiles upon all our efforts and stimulates us to greater endeavours. The fault is altogether ours if we do not regard these portents. The beneficent Creator, having established these celestial bodies " for signs and for seasons," is ever faithful. He puts up the signals on every occasion. It is for us to apprehend and read them.

In King Edward VII.'s horoscope we have the attachment which led to his marriage indicated by

Venus conjunction Moon mundo	19° 25'
Moon conjunction Venus con.	20° 7'

The attempt on his life by the maniac Sipido, when as King he was travelling in Germany, measures to an arc of 68° 25', and the appropriate direction was—

Sun opposition Neptune zod.	58° 21'

ILLUSTRATION

The death of the Empress Frederick (Princess Royal) in August 1901 was indicated by the direction—

Midheaven conjunction Saturn 59° 43'

The death of his royal mother, Queen Victoria, requires an arc of 59° 14', and we find the appropriate directions—

Midheaven square Moon zod. 58° 58'
Ascendant opposition Moon 59° 19'
Saturn semisq. Ascendant 59° 22'
Midheaven conjunction Saturn zod. 59° 42'

These illustrations will doubtless serve for all practical purposes, and they can be worked out at leisure by those who wish to exercise themselves in this art.

Other methods than that here illustrated must claim our attention, inasmuch as they have consistently been advocated by various authors. There are, moreover, several points which may be considered as debateable, and these also have to be considered before our work is rendered complete. We must therefore pass on.

CHAPTER X

PTOLEMY AND PLACIDUS

IT is generally conceded that the system of directing which has so far occupied our attention first originated as a measure of time in the mind of Claudius Ptolemy, the famous geographer, mathematician and astronomer of Alexandria, who flourished in the second century of our era, and wrote a standard work on the subject of astrology called in the Greek *Tetrabiblos*, and in the Latin *Quadripartite*, being four books on the Influence of the Stars. He also wrote the *Syntaxis* and the *Almagest*, which, together with his work on astrology, have been translated into every language in Europe and into many Oriental languages also.

From the writings of Sir Isaac Newton we have evidence that there were many sources of information open to Ptolemy in the pursuit of astrological knowledge, and there is no reason to suppose that he did not avail himself of them fully, for none has ever suggested that astrology as a science was first promulgated by him. But it may certainly be affirmed that Ptolemy gave to the Western world the first scientific exposition of the subject. There are two Latin editions of the work and one in Greek. The best translation that we have is the paraphrase of Proclus from the Greek text rendered into English with extensive commentary by J. M. Ashmand, and recently published as a supplement to *Coming Events*. Ashmand has followed the Elzevir text, dated 1635.

The name of Claudius Ptolemy will be revered wherever astronomy and astrology are studied. It is enough for the purpose of this sketch to note that he was born at Pelusium in Egypt, and became a brilliant disciple of the Alexandrian School. It appears that he was born about the year 80 A.D., flourished during the reigns of Adrian and Antoninus Pius, and died in the seventy-eighth year of his age.

Of Placidus de Titus, who first rendered a studied version of Ptolemy's work on astrology, we have very little information. It appears that he was known as Didacus Placidus, and was a native of Bologna, became a monk, and was appointed mathematician to the Archduke Leopold William of Austria. He wrote in the early part of the seventeenth century a work entitled the *Primum Mobile,* in which he gives a thorough digest of the teaching of Ptolemy. The best English translation is by Cooper. Placidus showed that Ptolemy recognised two sets of directions arising out of two sets of planetary positions, one in the zodiac and the other in the world, i.e. in the prime vertical. To Placidus remains the credit of having elaborated that part of directional astrology which has regard to directions in mundo.

Ptolemy makes it clear in his chapter on the " Number of the Modes of Prorogation " (bk. iii., ch. xiv.) that " when the vital prerogative is vested in the Ascendant, the anareta or killing planet may be brought to it by oblique ascension ; and if it be vested in the Midheaven or a body there situate, then direction is to be made by right ascension. If on the occidental horizon, the degrees of oblique descension are to be reckoned. But if not in either of these three places, but in some intermediate station, it should be observed that other times will bring the succeeding place to the preceding one, and not the times of ascension or descension nor of meridian transit as already declared.

" For, if it be desired to calculate agreeably to nature, every process of calculation that can be adopted must be directed to the attainment of one object—that is to say, to ascertain in how many equatorial times the place of the succeeding body or degree will arrive at the position preoccupied at the birth by the preceding body or degree, and, as equatorial times transit equally both the horizon and the meridian, the places in question must be considered in regard to *their proportionate distances from both these, each equatorial degree being taken to signify one year."*

Here Ptolemy makes it clear that he directs a body in the heavens to one that precedes it, or a body to a degree that precedes it, which direction is formed by the diurnal rotation of the Earth on its axis from west to east. He also makes it clear that he uses the proportionate distances of bodies from both the horizon and meridian as the basis of the calculation, and the arc of direction is the intervening degrees (equatorial) between them, at the rate of one equatorial degree for a year of life.

It is evident, therefore, that he takes a proportion of the semiarcs, or, as he calls them, " the horary times," of the planets involved. These arcs he describes as parallel to one another and to the Equator, but cutting the circle of the horizon at various degrees of obliquity.

Obviously, therefore, we have to take proportion of their semiarcs and meridian distances, exactly as we have been instructed in the foregoing exposition ; and as these semiarcs are regulated by the latitude of the place of birth and the corresponding ascensional differences of the planets, the positions of the bodies will have respect to the prime vertical and will be their apparent places in the plane of that circle. But it is important to note that Ptolemy says nothing concerning converse directions, whether in mundo or in the zodiac.

That he recognises the mundane position of a body as distinguished from the apparent place of its " degree " of longitude is obvious from his mentioning both in the same sentence ; and we distinguish ourselves between the mundane and zodiacal conjunctions only by reference to the body of the planet in the first instance and its longitude in the other case.

To Claudius Ptolemy, therefore, may rightly be accorded the honour of having set astrologers upon the right track with regard to the correct measure of time by reference to the equatorial degrees separating one body from another, or one body from the longitude or aspect of another, as seen from the place of birth.

There is little doubt, from the illustrations of his method that Ptolemy gives, that he uses the " ascensional " times in all cases due to the latitude of the place of birth ; and this method serves very well not only for directions to the Ascendant and Descendant, but also for intermediate positions when the planets are in the same or different quarters and on the same side of the meridian, for then their arcs may be measured with great facility and approximate accuracy from the Tables of Houses alone.

Illustration

1. Bring the Sun to the place of Mars in the horoscope of Ruskin.

	h.	m.
The sidereal time *(S.T.)* on the Midheaven when Mars' place rises is	15	49
That when the Sun rises is	16	44
Difference in R.A. on the Midheaven in S.T.	0	55

Divided by 4, this gives 13° 45' as the arc of direction.

The same arc of direction when exactly calculated by the semiarc method is 13° 49'.

2. Bring the Sun to the conjunction with Venus in zodiac.

	h.	m.
The S.T. *sidereal time* at sunrise (as above) is	16	44
That when Venus' place rises is	14	35
- Difference	2	9

This gives an arc of 32° 15'.

3. Bring Saturn to the place of Sun in zodiac. The declination of Saturn is 6° 54' S., and this answers to the longitude of Pisces, 12° 37'.

	h.	m.
sidereal time S.T. on Midheaven when this point rises	17	30
S.T. on Midheaven when Sun rises	16	44
Difference	0	46

This gives an arc of 11° 30'.

4. Bring the Moon to the opposition of Venue in zodiac.

The declination of the Moon is 25° 39', which exceeds that of any degree of the zodiac owing to the Moon's extreme latitude north added to the declination of its longitude. But reference to the Tables of Ascensional Difference and Right Ascension will show that its oblique descension answers to the twelfth degree of the sign Leo, which is the same as the oblique ascension of Aquarius 12°. Then the arc between the place and Venus in zodiac and Aquarius 12° will be the arc of direction. Thus :

		h.	m.
S.T. on Midheaven when Venus long. rises		14	35
S.T. on Midheaven when the 12th of			
Aquarius rises		16	30
	Difference	1	55

This gives an arc of 28° 45'.

5. Bring the Sun to the opposition of Uranus in zodiac.

Take the opposite degree of the zodiac to that held by Uranus, and bring the Sun to it by oblique arc.

		h.	m.
S.T. when Gemini 23° 25' rises		21	21
S.T. when Sun rises		16	44
	Difference	4	37

This gives an arc of 69° 15'.

6. Bring Sun to par. Uranus in zodiac direct.

The declination of Uranus is 23° 24', which answers to that of Cancer 4°. Find the arc between this and the Sun.

		h.	m.
S.T. on Midheaven when Cancer 4° rises		22	06
S.T. on Midheaven when Sun rises			
in Aquarius 18° 45'		16	44
	Difference	5	22

This gives an arc of 80° 30'.

These examples will serve to show that without recourse to the elaborations of a speculum or the use of proportional logarithms in the computation of proportional arcs, Ptolemy could, by the mere use of a table of ascensions under any latitude, find the time of an indicated event within an arc of 30' and even less, which, having regard to the approximations which are frequently adduced as " arcs for the event " when both are accurately known, show that they would serve for all practical purposes. I most frequently calculate arcs of direction in this manner, bringing out the results to the nearest quarter of a degree, which measures to three months of time. Ptolemy had constructed such tables, as appears from his *Almagest,* and this is obviously the method he used. In other words, he recognised no other directions than those that could be calculated by the difference of the oblique ascensions of the planets and of their longitudes, taking the oblique ascension of their opposite degrees when the arc was formed by descension of a body.

A table of oblique ascensions such as that published by Worsdale enables the calculation to be made with even closer exactness. It has only to be remembered that when we are directing the body of a planet to the body or longitude of another, the longitude corresponding to its declination must be dealt with, and not the longitude of the body itself, as the above examples will sufficiently indicate.

CHAPTER XI

DIRECTIONS UNDER POLES

THIS method has been much advocated, and especially by Mr R. C. Smith, the first of the almanac writers under the pen-name of " Raphael." It consists in directing a significator under its own pole instead of under the pole of the place for which the horoscope is cast.

To find the Pole of a Planet

Take its R.A., declination, and semiarc.
Then say :

As the semiarc is to 90°,
So is its meridian distance
To the difference of its circle of position and
   the meridian.

And this difference, compared with its meridian distance, will give its ascensional difference under its own pole.

Then having this and also its declination, from the sine of its ascensional difference under its own pole take the tangent of its declination, and the remainder will be the tangent of its pole.

Example.—In the horoscope of Ruskin find the pole of the Sun.

The R.A. of Sun is 321° 12', the meridian distance (below) 108° 44', the semiarc 110° 1', and the declination 15° 13'.

DIRECTIONS UNDER POLES

Semiarc 110° 1'	prop. log.	0.21381
Arith. comp.		9.78619
Quadrant of 90°		0.30103
Meridian distance	108° 44'	0.21891
Difference	88° 57'	0.30613

Asc. diff. under pole	19° 47'	log. sine	9.52951
Sun's declin.	15° 13'	log. tang.	9.43458
Pole of Sun =	51° 13'	log. tang.	10.09493

It is thus seen that the pole is measured along the tangent by its distance from the meridian or nadir, according as the body may be above or below the Earth at the time. At the meridian the pole would be 0, and at the horizon it would be the same as the latitude. Here " pole " is the same as polar elevation. The difference 88° 57' indicates the place of the circle of position from the plane of the meridian circle. Circles of position are small circles which are parallel to the great circle of the meridian and at right angles to the great circle of the horizon. They are like lateral circles of latitude in relation to which the meridian stands as equator and the Ascendant and Descendant as poles. Hence, if a planet be on the cusp of a house, it will have the same pole as that house.

Having calculated the poles of all the planets, and of the Sun and Moon, direction of one to another of them is thus made.

Rule.—Take the oblique ascension (or descension, as the case may be) of the promittor or body directed to under the pole of significator, and the difference of this from the oblique ascension (or descension) of the significator under the same pole is the arc of direction.

To find the oblique ascension of a body under the pole of another directed to it, to the log. tang. of its declination add the log. tang. of pole of the body directed, and the sum will be the log. sine of its ascensional difference under that pole. From this its oblique ascension can be found by referring it to its R.A. according to the rule (see " Definitions," Chapter I.).

Example.—Direct the Sun in the example horoscope to the place of Venus in the zodiac.

The declination of Capricorn 5° 49' is 23° 20'. The Sun's pole is 51° 13'. Then—

Pole of Sun, 51° 13'	log. tang.	10.09493
Dec. Venus long.	log. tang.	9.63484

Asc. diff. of aspect	32° 28' log. sine	9.72977
R.A. of aspect .	276° 20'	

O.A. of aspect	308° 48' under pole of Sun.
O.A. of Sun	340° 59' under its own pole.

Diff.	32° 11' =arc of direction.

Applying this method to the hint I have already given as to the use of tables of oblique ascension, or tables of houses for various latitudes, we can calculate this arc perfectly well with a table of the houses for latitude 51° 13', which is the pole of the Sun. And we can calculate all the solar arcs by this means from the same table. Then if we find the pole of the Moon, and refer to the Table of Houses for equivalent latitude, we shall be able to take out all the directions of the Moon under its own pole. The directions of the Ascendant will, of course, be made under the pole of the place of birth, and those of the Midheaven by right ascension only. So that what appears at first a complex and exhaustive piece of work can readily be done by tables of houses, or tables of oblique ascension for various latitudes, in next to no time, as the saying is. And this, I think, may be adjudged the most popular contribution to the theory and practice of primary directions that I have been able to make.

Example.—Direct the Sun under its own pole to the opposition of Uranus in the zodiac.

The Sun's pole is 51° 13'. Therefore take in hand the Tables of Houses or the Tables of Oblique Ascension for latitude 51° 13' N.

DIRECTIONS UNDER POLES

The opposition of Uranus falls in Gemini 23° 25'.

	h.	m.
S.T. on Midheaven when		
Gemini 23° 25' rises	21	21
S.T. when Sun's place rises	16	43
	4	38

This converted into arc of R.A. =69° 30' =arc of direction.

Example.—Direct the Sun under its own pole to Venus in the zodiac. Pole of Sun =51° 13'.

	h.	m.
S.T. on Midheaven with Sun rising	16	43
S.T. on Midheaven with		
Capricorn 5° 49' rising	14	35
Arc of direction, Sun conj. Venus		
zodiac =difference	2	8

This is equivalent to 32° 0'.

By exact calculation we found it formerly to be 32° 11'. The difference is inconsiderable from the point of view of probable time of the event.

As to the merits and demerits of these divergent systems of directing, I leave my readers to decide for themselves. *Expertientia docet.* I hold no brief for either system, my business being merely to represent and to simplify. This I think I may claim to have done.

The system of directing under the semiarcs in the prime vertical is that which was followed by Ptolemy. The system of directing under the poles of the planets is of considerably more recent origin, and dates to the seventeenth century only. It consists, as will be seen, in directing in the circle of observation due to the pole of the significator or planet directed. The difference is that which one may note as between the tables of houses for one latitude and another. Nothing is simpler or more demonstrable. I leave it at that.

But in general practice it will be found that equally close results may be obtained by simple proportion and the use of the tables. Take the following hint for what it is worth. I am quite satisfied in my own mind that what we call primary directions seldom or never operate exactly to time, and if we correct the observed time of birth by one direction for an event we shall find that subsequent directions are not on schedule time. We have to allow a latitude for the operation of these directions. Such being the case, and, in the experience of the best artists, the import of primary directions being accelerated or retarded by the secondary directions and transits, we do not need to observe scruples. Approximations are always valuable.

The following may be regarded as the *via loetitia* in primary directing:—

Rule 1.—As the semiarc of the planet whose pole is required is to 90° of the prime vertical, so is the distance of the body in right ascension from the meridian (upper or lower as the case may be) to its proportional distance in the prime vertical.

Rule 2.—From the sine of their difference subtract the tangent of the planet's declination. The remainder is the tangent of its pole.

Rule 3.—For all directions under the pole of that planet or significator use the Tables of Houses for that latitude which answers to its pole.

Rule 4.—Find the difference between the ascension of the body (by sidereal time or right ascension on the Midheaven) and that of the planet directed to. This will be the arc of direction.

Note.—If the planets involved or the positions involved are between the tenth and fourth westward, take the ascensional degrees of the opposite places.

Rule 5.—Direct the Midheaven by right ascension only, and the Ascendant by oblique ascension under the latitude of birth. Direct the Sun under its own pole and the Moon under its own pole. This completes the entire scheme of primary directing.

Example.—In the horoscope of Ruskin the Sun was found to have a pole equal to the latitude of 51° 13' N. (see p. 61). It must therefore be directed under the Ascendant of 51° 13'. Similarly, the Moon, whose pole is 47° 27', must be directed under the latitude of that degree. A significator on the Midheaven would thus be directed by right ascension only, as stated by Ptolemy (see p. 55).

For directions of the Sun to other bodies, therefore, we use the Tables of Houses for 51° 13'. Those for Taunton are 51° 1', which is deemed near enough.

1. Direct the Sun to Jupiter in the horoscope.

	h.	m.
S.T. on Midheaven with Sun rising	16	41
S.T. on Midheaven with Jupiter's long. rising	15	55
Arc of direction 11° 30', equivalent to S.T.	0	46

2. Direct the Sun to Mars.

	h.	m.
Sun rising as before, S.T. on Midheaven	16	41
Mars rising, S.T. on Midheaven	15	43
Arc of direction = 14° 30'	0	58

3. Direct the Sun to Mercury in zodiac.

	h.	m.
Sun's rising as before	16	41
Mercury's longitude rising	15	39
Arc of direction = 50° 30'	1	2

4. Direct the Sun to Venus' longitude.

	h.	m.
Sun's rising as above	16	41
Place of Venus rising	14	30
Arc of direction = 32° 45'	2	11

5. Direct the Sun to Neptune in zodiac.　　h. m.
　Sun's rising as before　　　　　　　　　16 41
　Neptune's long. rising　　　　　　　　　13 59

　Arc of direction = 40° 30'　　　　　　　2 42

6. Direct the Sun to Uranus in zodiac.　　h. m.
　Sun's rising as above　　　　　　　　　16 41
　Uranus' long, rising　　　　　　　　　　13 36

　Arc of direction = 46° 15'　　　　　　　3　5

7. Direct the Sun to opposition of Moon in zodiac.
　Sun's rising as before　　　　　　　　　16 41
　Rising of Capricorn 27° 8', S.T.　　　　15 47

　Arc of direction = 13° 30'　　　　　　　0 54

The various aspects to these promittors can be picked up *en route* as we bring the Sun from the horizon to the Midheaven, which it reaches in an arc of 69° 59'=70 years nearly.

We cannot direct Sun to Saturn by the diurnal motion of the Earth, and so we must bring Saturn up to the Sun's place. This involves knowing the pole of Saturn.

We may also bring Saturn to the Ascendant under its own pole. But if we were to bring the Sun to Saturn under the Sun's pole, that would be a *prenatal direction,* for the Sun cannot go back from the position it has attained and sink below the eastern horizon. We have therefore no alternative but to regard these directions as invalid, or to admit the thesis already suggested, that in these directions, made contrary to the apparent motion of the bodies in the heavens, we are dealing with the localised impress of the planet at the moment of birth, which impress is carried by the Earth up the western heavens and down the eastern heavens, so that the Sun's localised imprint is here carried down to the place of Saturn. And this is conformable to the theory of directions under the poles of the significators.

CHAPTER XII

THE PART OF FORTUNE

FOR a considerable time there was much discussion as to the correct method of finding the place of the Part of Fortune. This, it should be explained, is one of the old Arabic points, which, like the Pomegranate, the Sword, and others, were regulated by the distances of the several bodies from the Sun in the zodiac, the particular point referred to being the same distance in zodiacal degrees from the Ascendant.

It was when astrologers came to apply this theory to the system of primary directions in vogue that the trouble arose as to the correct method of computing this point.

I think that the easiest expression of the case is this :— the Part of Fortune is a mundane point answering to the distance of the Moon from the Sun in the zodiac. Thus in the horoscope of Ruskin the Moon wants 21° 38' from the opposition of the Sun, and therefore the Part of Fortune will be 21° 38' below the western horizon in mundo. Its mundane position will therefore be 8° 22' inside the 6th house.

Its meridian distance will be 68° 22', and its pole 39° 13'. Under this pole we may direct it to aspects in the zodiac, and in mundo. It has been suggested that the Part of Fortune cannot be directed, but can only receive directions from other significators and the planets. This is surely nonsense. Any point in the heavens having been defined and located is carried by the motion of the Earth on its axis from its radical place to others successively in a direction that is contrary to the rotation of the Earth. Hence the Part of Fortune will here be carried down the heavens from the 6th to the 5th and from that to the 4th house successively, forming both mundane and zodiacal aspects under its own pole.

The pole of the Part of Fortune and that of Saturn being near to one another, they must be near a mundane parallel, on the same side of the horizon.

There are, however, other suggested methods of taking the place of the Part of Fortune.

Ptolemy says (bk. iii., ch. xii.): " The Part of Fortune is ascertained by computing the number of degrees between the Sun and Moon, and it is placed at an equal number of degrees from the Ascendant in the order of the signs. It is in all cases, both by day and night, to be computed and set down, that the Moon may hold with it the same relation as that which the Sun may hold with the Ascendant ; and it thus becomes, as it were, a lunar horoscope or Ascendant."

It is therefore clear that Ptolemy intended degrees of oblique ascension or descension, and not merely degrees in the zodiac, the relations of which, in regard to the horizon of any place, are continually changing.

Thus in the horoscope of Ruskin we have—

O.A. of Sun		341° 13'
O.D. of Moon	157° 26'	
add	180° 0'	337° 26'
		3° 47' Moon to
		oppos. Sun.
O.D. of 7th		159° 56'
		156° 9' O.D. of
		Part of Fortune.

This gives us a position answering to the 10th degree of Leo, and therefore close to the Moon.

Placidus says : " Let the Sun's oblique ascension taken in the Ascendant be subtracted always from the oblique ascension of the Ascendant, as well in the day as in the night, and the remaining difference be added to the Moon's right ascension ; the sum will be the right ascension of the Part of Fortune, which will have the Moon's declination."

THE PART OF FORTUNE

In the example horoscope the oblique ascension of the Ascendant is 339° 56', from which take the Sun's oblique ascension 341° 13' (adding 360 for subtraction), and the remainder is 358° 43', which add to the right ascension of the Moon 120° 17', and the sum is the right ascension of the Part of Fortune 119° 0'.

The right ascension of the *imum coeli* being 69° 56', the meridian distance of the Part of Fortune will be 49° 4', and its semiarc will be that of the Moon, 52° 51', as it has the same declination as the Moon in all cases. Then semiarc 52° 51' — 49° 4' = 3° 47', which is the same as we derived from the method of Ptolemy. For there we found the oblique descension of the Part of Fortune to be 156° 9'; and the oblique descension of the 7th being 159° 56', the difference is 3° 47'.

The method of Placidus appears preferable in that we derive at once the right ascension and meridian distance of the Part of Fortune.

The question is, however, whether either is true, and only directions made by the position as thus derived can settle the point in debate.

To enable the student to at once work out the primary arcs, we here append the speculum in the example horoscope:

R.A.	Ner. Dist.	Semiarc.	Horiz. Arc.	Cusp. Dist.
119° 0'	49° 4'	52° 51'	3° 47'	3° 47'

These elements at once suggest that the pole of the Part of Fortune can be found, and direction made by the Part of Fortune in mundo and zodiac to the planets, just as if it were a definite body.

The fact that it is merely a symbol, a point in space, does not in the least invalidate its significance in human affairs, as some impulsive students have suggested. For what else are the degrees of the zodiac known as the Midheaven and Ascendant ? They are points in space which bear a definite relationship to a particular place at a given time. They do not need to be identified with a star in the heavens in order to obtain a significance in the horoscope. Every tyro in astrology knows as an absolute fact that these points have a very demonstrable significance in a horoscope, and that transits of the major planets over these points, and the passing of these points by planets in direction, are attended by events which leave no shadow of doubt that they are an essential part of the signalling apparatus by which we are forewarned of coming events. And if these, why not the Part of Fortune ? Call it a " myth " if you like, but understand that a myth is a " veil " designed to hide a truth which a symbol is said to embody. The symbol handed down to us is identical with that used in China and also in Egypt to indicate " land, territory, a field."

CHAPTER XIII

LUNAR PARALLAX AND SEMI-DIAMETER

AMONG the problems modernly confronting the student of directional astrology, that of the horizontal parallax of the Moon is perhaps one of the most important and at the same time most perplexing.

The places of the planets as indicated in the horoscope are the geocentric longitudes. They are computed from the standpoint of an observer. But as the place of observation is on the surface of the Earth and not at its centre, the observed position of the Moon will not exactly coincide with its computed geocentric longitude. In the case of the Sun and planets, the distances from the Earth are so great as to render the parallax inconsiderable, that of the Sun being only 9", and the parallaxes of other bodies beyond it being proportionately less. But in regard to the Moon, its nearness to the Earth renders its parallax of importance if we are to regard the Moon as affecting us by its direct ray. The nearer the Earth it may be, the greater is the angle of parallax. It is therefore greatest at the perigee and least at the apogee of the Moon.

As the amount of parallax depends on the Moon's place in its orbit, we make use of the apogee as a point of departure, and the Moon's distance from that point in its orbit where it is furthest from the Earth is called its anomaly.

By comparing the calculated place with the observed place it has been found that the difference of the two at the apogee is 53' 53", and at perigee 61' 23". It will be sufficient for our purpose if we call these 54' and 61' respectively. By the use of the " Ready Reckoner " the amount of the anomaly can always be found for any date or hour, and the corre-

71

sponding parallax is set against it. The table is here repeated for convenience.

TABLE OF ANOMALY.

Epoch 1800, Jan. O^d O^h O^m = 9^s 20° 20'.

Years.	Add.			Days.	Add.			Anom.		Hor Par.
	s	°	'		s	°	'	s	°	'
1	2	28	43	1	0	13	4	0	0	54
2	5	27	27	2	0	26	8		6	55
3	8	26	10	3	1	9	12		12	55
4	0	7	57	4	1	22	16		18	55
5	3	6	40	5	2	5	19		24	56
6	6	5	24	6	2	18	23	1	0	55
7	9	4	7	7	3	1	27		6	55
8	0	15	54	8	3	14	31		12	55
9	3	14	38	9	3	27	35		18	55
10	6	13	21	10	4	10	39		24	56
20	1	9	46	11	4	23	43	2	0	56
40	2	19	32	12	5	6	47		6	56
50	9	2	53	13	5	19	51		12	56
60	3	29	18	14	6	2	55		18	57
70	10	12	39	15	6	15	58		24	57
80	5	9	3	16	6	29	2	3	0	57
90	11	22	24	17	7	12	6		6	58
100	6	18	49	18	7	25	10		12	58

Months.	Add.			Days.	Add.			Anom.		Hor Par.
				19	8	8	14	18		59
				20	8	21	18	24		59
January	0	0	0	21	9	4	22	4	0	59
February	1	15	1	22	9	17	26		6	59
March	1	20	50							
April	3	5	51	23	10	0	30		12	60
May	4	7	48	24	10	13	34		18	60
June	5	22	49	25	10	26	37		24	60
				26	11	9	41	5	0	60
July	6	24	46							
August	8	9	47	27	11	22	45		6	60
September	9	24	48	28	0	5	49		12	60
October	10	26	45	29	0	18	53		18	61
November	0	11	45	30	1	1	57		24	61
December	1	13	42	31	1	15	1	6	0	61

LUNAR PARALLAX AND SEMI-DIAMETER

Example.—Find the Moon's anomaly for 8th February 1819, and the corresponding horizontal parallax.

		s	o	,
Epoch 1800		9	20	20
Add 19		9	27	59
February		1	15	1
8 days		3	14	31
Anomaly	=	0	17	51

The Moon is therefore within 18° of its apogee or furthest distance from the Earth, and its parallax will therefore be near its minimum. Our table shows that the parallax due to this anomaly is 55'. This would be the difference between the Moon's geocentric longitude and its observed position from the surface of the Earth if it were exactly on the horizon. At the meridian the parallax is 0, and at the horizon it differs, as stated, from 54' to 61' according to the distance of the Moon from the Earth, *i.e.* its place in its orbit.

Now, as the horizon is at all points 90° from the zenith or nadir, we can make one of these the apex of a triangle, of which the zenith distance of the Moon at transit is the perpendicular and the base its meridian distance. From these we may find the hypotenuse, which will be the Moon's zenith distance at the time of birth.

Thus, in the example horoscope the latitude of the place is 51° 30' N., and the Moon has latitude 5° 1' N., which therefore must be subtracted, leaving 46° 29' as the zenith distance of the Moon at transit of the nadir. Its meridian distance is found from the speculum to be 50° 21'. Then

Log. cosine 50° 21'	9.80489
Log. cosine 46° 29'	9.83794
Log. cosine 63° 52'	9.64283

And as 90° is to 55', so is 63° 52' to 39', which is the Moon's parallax, and by which amount she is apparently depressed further below the horizon than she is computed to be. This will affect its meridian distance, etc. The directions of the Moon, if operating dynamically by right lines of energy upon any part of the Earth instead of via the centre

73

of the Earth, will hence be affected ; and it remains a problem worth some close study and consideration as to what view ought to be taken. It is sufficient here to have indicated the method of calculation. It is one of the factors in the vexed problem of " the uncertain Moon," which has frequently been charged with an inconstancy altogether absent from the directions of the Sun and planets.

The semi-diameters of the Sun and Moon have often been resorted to in order to accommodate a directional arc to the date of an event. Allowing, as is undoubtedly the case, that primary directions have an orb of influence within the limits of which it may be said they begin to operate, attain their maximum, and pass off, there yet remains the fact that one would naturally expect the maximum to coincide with the most marked phase of a crisis in the life. This appears to be acknowledged, inasmuch as practitioners in the art of directing make use of arcs of direction, measured from the centres of bodies as determined by their longitudes, in order to correct approximate times of birth. This correction can only be legitimately made on the supposition that arcs of direction are close, if not exact, to the time of the events they are held to signify.

And unless there were this fundamental integrity of the system of direction advocated, unless there was a close agreement throughout a life between the arcs of direction and the events portrayed, there would be no use in making the calculations.

Our longitudes are geocentric and apply to the apparent centre of the bodies. The apparent diameter of the larger planets, on account of their great distance from the Earth, is inconsiderable. But when we come to the Sun and Moon, which are the chief significators, and the bodies that are directed to form the aspects of the promittors, we are concerned with orbs that have a visible diameter. The Sun on account of its immense size, and the Moon on account of its close proximity, appear to have a diameter of about half a degree, or from the centre to the limb about 15'. This becomes an important consideration when we are directing either of them to the aspect or conjunction of one of the planets, inasmuch as from first to last contact of the disc of the luminary with the said planet or aspect there will be an included arc of half a degree, and this means six months of time according to the Ptolemaic measure of 1° = 1 year. Hence it may well be that a direction is increscent for three months before it attains its actual centrality and maximum strength, and another three months may transpire before the effects wear off. And if to this we add the fact that directions formed at the tropics, *i.e.* near Cancer or Capricorn 0, are very slow in formation (as may be seen from the Tables of Declination), 4° of longitude including only 1' of declination, it will readily be understood that there is ample room for " latitude " in the timing of events.

It seems desirable, therefore, that a few cases of very well-observed birth-times should be taken, and the arcs of direction computed very closely ; and then that these arcs should be compared with the course of events, so that an estimate of the value of the semi-diameters of the Sun and Moon may be made.

The apparent semi-diameter of the Moon is controlled by the same factor as the parallax, namely, its place in the orbit and consequent distance from the Earth. It may be useful to mention that the semi-diameter of the Moon is approximately twenty-seven one-hundredths of the parallax. Therefore multiply the parallax by 27 and divide by 100. Thus, when the parallax is 54', the semi-diameter of the Moon is $54 \times 27 \div 100 = 14\frac{1}{2}'$, and when the parallax is 60, the semi-diameter is $60 \times 27 \div 100 = 16'$.

The Moon directed to the opposition of the Ascendant in the example horoscope works out at 2° 30' ; but as the horizontal parallax of the Moon is 55', its semi-diameter will be nearly 15', and therefore the direction would read :

Asc. oppos. Moon in mundo,	first contact	2° 15'
" "	middle	2° 30'
" "	last contact	2° 45'

thus giving a possible range of 30', or six months for the duration of this indication. This may help to account for the variability that has been noticed in regard to lunar directions, and possibly we may also have to consider taking the parallax into account. The solar directions will be affected by semi-diameter of the Sun, but not appreciably by parallax.

CHAPTER XIV

LUNAR EQUATIONS

UNDER this head I propose to examine a problem of some interest which appears to have escaped general recognition, but which may very well be considered with the questions of parallax and semi-diameter as having some connection with the noted irregularity of primary lunar directions.

Take an illustration from the horoscope already submitted. We would direct the Moon to conjunction with the nadir, which direction is known as " Midheaven opposition Moon in mundo." It is measured by the arc of the Moon's meridian distance, 50° 21', and is formed by the rotation of the Earth on its axis, by which the Moon is carried down the western heavens until it makes its meridian transit.

The theory underlying this direction is that there is a permanent significance and value attaching to the radical positions of the Midheaven, Ascendant, and other significators, which is unaffected by the subsequent changes taking place amongst the planets, either on account of their apparent motions in the heavens or their real motions in the zodiac. But we have now to consider whether there may not be some value attaching to these subsequent motions of the bodies in the zodiac. These motions, within the narrow limits of time comprised in the formation of directions in a life of ordinary length, would not be appreciable in the case of the planets or the Sun, but in the case of the Moon there would be a quite appreciable increment owing to the velocity of that body in its orbit.

Thus the arc of 50° 21' cited above would occupy the interval of 3 hours 25 minutes, during which the Moon will have increased its longitude by about 1° 42', so that it would not actually make the meridian transit for another 7 minutes, although its radical place would then be exactly on the nadir. Its right ascension will be increased by about the same amount, and therefore the actual arc of direction from the time of birth until the bodily transit of the nadir would be about 52° 3'. So far as this case is concerned it is worthy of notice that this arc of the second distance of the Moon to the opposition of the Midheaven, and therefore to the mundane square of the Ascendant, coincided exactly with a period of serious illness and trouble in the life of Ruskin, whereas the arc M.C. opposition Moon in mundo, 50° 21', exactly coincided with the election of Ruskin to the Slade Professorship of Fine Art, a distinction which brought him into the highest position in his sphere of life. Obviously, therefore, the second distance of the Moon is by far the most appropriate. Let us look at another direction from the same point of view. Direct the Moon under its own pole to the opposition of Saturn.

The Moon's pole is 47° 27', and its ascensional difference under that pole, derived in the process of finding the pole, is

	31°	32'
Its right ascension	120°	17'
Its oblique descension under its pole	151°	49'
Add	180°	0'
Oblique ascension of opposition Moon=	331°	49'

Then for Saturn's oblique ascension under the same pole—

Pole of Moon tang.	10.03712	
Tang. Saturn's decl.	9.08283	
Ascl. diff. Saturn sine	9.11995 =	7° 31'
R.A. of Saturn		348° 54'
O.A. of Saturn		356° 25' under Moon's pole.
O.A. of Moon's oppos.		331° 49'
Arc. of Moon oppos. Saturn=		24° 36'

This corresponds with Ruskin's leap into public estimation and fame, for which we have the arc of direction Sun sextile Midheaven in mundo. Most certainly the Moon to opposition Saturn could not be regarded as in the least degree akin to the nature of events then current in the life of the great artist.

But this arc took 1 h. 38m. 24s. to complete, and during that time the Moon had increased its R.A. by some 49' ; and as we are bringing Saturn up to the opposition of the Moon under the pole of the Moon, we shall have to curtail the direction by 49', which results in an arc of 23° 47'. This is nearly a year in advance of Ruskin's great advent, and may very well have coincided with a period of stress and indisposition.

The Moon to the opposition of Venus comes into force at about thirty years of age, or in the thirtieth year of life, when he married ; but by adding the increment due to the time of direction to the radical place of the Moon we get an arc which falls out a whole year later, when it is certain Ruskin realised his disappointment.

The directions of the Sun during the course of sixty years would only be affected by an increment of 10', and they can always be relied upon ; but the directions of the Moon are at present very unsatisfactory, and it has been thought that this question of second distances may serve not only to indicate why lunar primary directions are inconstant, but why also they appear to have a more durable influence than those of the Sun. The suggestion is that from the time the direction is formed to the radical position of the Moon to the time that it is formed to the actual position of that body in the heavens, may be the extent of its duration ; and during this period, which naturally increases in length as the age increases, transits and other secondary indications may come up repeatedly to reinforce the portents of the lunar direction and bring them into play. Certain it is that there are many conditions affecting the directions of the Moon which arise out of its velocity, and to maintain its ancient reputation for inconstancy and fickleness it appears to have jealously guarded its secret even from the lynx eye of the practical astrologer. Whether we have succeeded in compassing the fickle goddess by this exposition remains to be decided by constant experiment conducted by several independent workers. In the cause of a scientific astrology this is worth carrying out, and it is to be hoped that qualified and unprejudiced students will communicate their experience.

It may assist the average student to know that all directions of the Moon to *succeedent* places will fall out sooner, while those to *precedent* places will fall out later, than indicated by the radical or first distance of the Moon, and the arc of direction must therefore be increased or decreased at the rate of 2' for every degree of the arc of direction. Thus an arc of 39° 15' requires 1° 18½.

CHAPTER XV

CUSPAL DISTANCES

WHEN giving instructions as to the method of directing bodies to aspects of the Ascendant and Midheaven in mundo, it is customary to affirm that one-third of a planet's semiarc is equal to a house-space, so that a planet that is one-third of its semi-arc above the horizon is held to be on the cusp of the 12th house, and when two-thirds of its semiarc above the horizon it is on the cusp of the 11th. But if this were actually the case, we should find that when on the cusp of a house the oblique ascension of an ascending planet is the same as the oblique ascension of the cusp of that house. Such is not the case.

Example.—Direct the Sun in Ruskin's horoscope to the sextile of the Midheaven in mundo. This aspect falls on the cusp of the 12th house.

The semiarc diurnal of the Sun is 69° 59', and one-third of this is 23° 20', to which add the Sun's distance under the horizon, 1° 17', and we get the arc of direction=24° 37'. The Sun is then on the cusp of the 12th house presumably. Let us see.

The R.A. of the Midheaven is 249° 56', to which if we add 60 we shall have the oblique ascension of the cusp of the 12th house, 309° 56'. Now, when the R.A. of the Midheaven is increased by an arc of 24° 37', the oblique ascension of the cusp of the 12th will be increased by the same amount, and will then be 334° 33', while the oblique ascension of the Sun is 341° 13'. Wherein lies the error ?

It lies in the fact that we are directing the Sun under the pole of the Ascendant, whereas we should direct it un-

81

der the pole of the 12th house cusp.

I here give a table of the polar elevation due to the various houses in several latitudes, from which, by proportion of their parts, we may derive the pole of any house for any minute of the included latitudes.

POLES OF HOUSES.

Lat.	Cusps of 3, 5, 9, 11.		Cusps of 2, 6, 8, 12.	
	°	'	°	'
45	18	57	34	11
46	19	37	35	10
47	20	19	36	10
48	21	2	37	10
49	21	46	38	12
50	22	33	39	15
51	23	21	40	19
52	24	12	41	24
53	25	5	42	31
54	26	1	43	39
55	26	59	44	48

The pole of the 12th house for the latitude 51° 30' N. is seen to be 40° 51', and if we direct the Sun under this pole we shall have the

Ascensional difference of Sun under pole of 12th	13°	36'
Right ascension of Sun	321°	12'
Oblique ascension of Sun under pole of 12th	334°	48'
Oblique ascension of cusp of the 12th house	309°	56'
Arc of direction	24°	52'

This, although not exact, is certainly nearer, and seems to justify the method of directing under the poles of planets.

CUSPAL DISTANCES

The fact, however, is that if we take a fixed pole for any house in a given latitude we shall always be in some degree of error, and for the simple reason that the semiarcs of the planets, being parallel to the equator, do not lie in the same plane as the prime vertical, which is the circle we divide into twelve equal parts to form the houses of the heavens. Therefore an equal division of the prime vertical will not result in an equal division of the semiarcs, and either we have to consider the poles of the houses as movable, or, as seems more consistent with the facts, we must regard the house-spaces as unequal. In other words, we shall find that the time (measured by degrees of R.A.) that the Sun remains in successive houses is unequal, and the same is to be said of any other body. When, therefore, we take one-third of the semiarc of a planet as equal to one house-space, we are indulging in a free use of the metaphysical concept that " all circles are equal to one another," as defined by the doctrine of Correspondences. Against this I have nothing to say except that it is not mathematics.

Now, just as we take the Sun's oblique ascension under the pole of the Ascendant in order to find its distance from the horizon, so we must take its oblique ascension under the pole of the 12th house in order to find its distance from the cusp of the 12th, and its oblique ascension under the pole of the 11th to find its distance from the cusp of the 11th. Its right distance from the cusp of the 10th will be its arc to that cusp, since the meridian has no polar elevation. Thus :

The pole of the Ascendant is	51° 30'
The pole of the 12th house	40° 51'
The pole of the 11th house	23° 46'

The Sun's declination is 15° 13', log. tang. 9.64380, and if to this we add the tangent of the poles of the houses successively we shall have the sine of the ascensional differences of the Sun under these poles, which, added to its right ascension, will give its oblique ascension under those poles. These are :

83

PRIMARY DIRECTIONS

O.A. of Sun under pole of 1st house	341° 13'
O.A. of Sun under pole of 12th house	334° 48'
O.A. of Sun under pole of 11th house	327° 59'
R.A. of Sun under pole of 10th house	321° 12'

Then, to find the arc of direction between the Sun and any of these cusps, we merely subtract the oblique ascension of the one from the other. The oblique ascensions of the cusps are :

Of the Ascendant	339° 56'
Of the 12th	309° 56'
Of the 11th	279° 56'
Of the Midheaven R.A.	249° 56'

Thus we have the following true arcs of direction of the Sun in mundo :

O.A. Sun under pole of Ascendant	341° 13'
O.A. of the Ascendant	<u>339° 56'</u>
Arc of Sun to conjunction Ascendant	1° 17'
O.A. of Sun under pole of 12th	334° 48'
O.A. of 12th house cusp .	<u>309° 56'</u>
Arc of Sun to sextile Midheaven mundo	24° 52'
O.A. of Sun under pole of 11th	327° 59'
O.A. of cusp of 11th .	<u>279° 56'</u>
Arc of Sun to sextile Ascendant mundo	48° 3'
R.A. of Sun under Meridian	321° 12'
R.A. of Midheaven	<u>249° 56'</u>
	71° 16'

And in all these cases the Sun will have the same oblique ascension as the cusp of the house to which it is directed, at the time of direction being completed. This is what we argue for and obtain.

Also we may find the degrees of R.A. which pass under the meridian while the Sun passes from the cusp of one house to the next, and thus the house-space of the Sun at its present declination.

As the whole diurnal arc of the Sun is less than 90, the house-space will be less than 30°.

Subtract the arc of direction of Sun conjunct Ascendant from the arc of direction Sun conjunct 12th = Sun sextile Midheaven. There remains 23° 35', the house-space of 12th house.

Subtract the direction of the Sun to the 12th from that to the 11th ; there remains 23° 11', the house-space of the Sun in the 11th. Subtract the arc of direction Sun cusp of the 11th from the Sun conjunct Midheaven ; there remains 23° 13', the house-space of the Sun in the 10th.

And the three house-spaces added together = 69° 59', which is the diurnal semiarc of the Sun.

Hence it appears that the mundane directions of planets must be taken in terms of the pole of the cusp to which they are directed. The cuspal distances of the planets must also be measured according to the same rule. This will affect all directions calculated by primary arcs on the semi-arc method now commonly in vogue.

But what appears of most vital importance as a legitimate conclusion drawn from this critique is that the correct method of directing to any body is by oblique ascension under the pole of that body, which is quite different from taking the direction under the pole of the body directed. At the same time, it appears to dispose of the semiarc method, except as a valuable approximation. For nothing can be more certain than that the cusps of the houses, measured in the prime vertical, are 30° distant from one another by oblique ascension.

These conclusions agree entirely with our mathematics, for we have seen that the house-space of the Sun in the 12th, due to its declination, is 23° 35' ; and if to this we add the Sun's direction (from below) to the Ascendant =1° 17', we have an arc of direction, Sun to conjunction cusp of 12th= Midheaven sextile

Sun in mundo, 24° 52', which is exactly what we found the direction of the Sun to be by oblique ascension when taken under the pole of the 12th house.

This proves, if anything can, not only that the correct method of directing is under the pole of the planet or position directed to, but also that the house-spaces are variable and depend on the several declinations of the planets, and thus on their oblique ascensions and descensions, taken under the poles of the successive houses.

By the semiarc method, taking one-third of a semiarc as equal to a house-space, we are dealing with an approximation which, although useful and facile, is not mathematically correct. Rather than that bad habits should become popular, I have undertaken a somewhat lengthy demonstration of this point, which I consider to be now settled beyond further debate.

CHAPTER XVI

SUGGESTED METHOD OF TRUE DIRECTING

As the result of this examination of the various methods of directing, both by semiarc proportions and by oblique ascensions under the poles, we may come to the conclusion that all the disparities which vitiate the present methods can be disposed of if we proceed along the lines to which our conclusions point. For this purpose we shall require a speculum containing :

1. The right ascension of a planet.
2. Its declination.
3. Its pole.
4. Its ascensional difference under its own pole.

The first of these will, of course, be worked as usual. The declination will be that given in the ephemeris. The pole of the planet will be that derived in the usual way from the ascensional difference of its proportional place in the prime vertical taken under its own declination, as already shown. Its cuspal distance will be the difference between its oblique ascension (or descension) taken under the pole of the cusp to which it is nearest and the oblique ascension of the cusp in the prime vertical. These are all the elements required for a complete calculation of all legitimate arcs of direction.

Directions must be made under the pole of the body to which we are directing another. The pole is the same as geographical latitude. It represents the latitude (geographical) or polar elevation (astronomical) at which the cusp of the house cuts into the circle of the prime vertical, or at which a circle of position cuts into it.

Thus in the following diagram let the great circle NZHS, etc., be the sphere of the Earth, of which N is the north pole, S the south pole. Also let ZN be the great circle of the prime vertical at an elevation from the Equator of 51° 30' N., and H–H the horizon intersecting it at right angles. Then HNZH will be the upper meridian and HSNH will be the lower meridian, the points Z and N marking the zenith and the nadir. The cusps of the 10th, 11th, and 12th houses are shown by the great circles cutting through the prime vertical at different elevations, and these answer exactly to the geographical latitudes (north) of the same values. Thus the pole of the Ascendant is 51° 30', that of the 12th, 40° 51', that of the 11th, 23° 46', and that of the 10th, 0° 0', as shown in the diagram, the ascensional difference being the arc in R.A. between N–S and H–H.

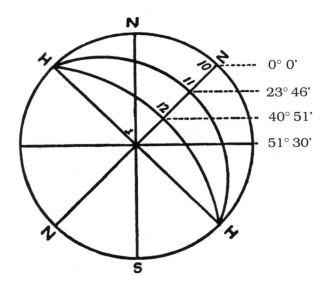

A circle of position is thus seen to be one which passes through a body and converges upon the horizon north and south exactly like an intermediate cusp of a house.

SUGGESTED METHOD OF TRUE DIRECTING

Rules for Directing

Rule 1.—Find the pole of the body or ecliptic position to which direction is to be made. Find the ascensional difference under this pole of the body to be directed. Apply this ascensional difference to the right ascension of the body to be directed, and obtain its oblique ascension (or descension, as the case may require) under the pole of the body to which direction is made. The difference between this and the oblique ascension of the body to which direction is made, taken under its own pole, will be the *arc of direction.*

Rule 2.—In mundane directions take the body of the promittor, i.e. body directed to. In zodiacal directions take its longitude.

Rule 3.—In directing to the aspect of a planet in mundo, its cuspal distance taken under its own pole must be directed to under the same pole.

This rule also serves for mundane parallels.

Here is the Speculum required for Ruskin's horoscope.

SPECULUM

Planet.	R.A.	Declin.	Pole.	Ascen. Diff.
Sun	321° 12′	15° 13′ 9.43458	51° 13′ 10.09493	19° 47′
Moon	120 17	25 39 9.68142	50 21 10.08147	35 24
Mercury	296 47	21 34 9.59688	44 55 9.99885	23 13
Venus	276 6	18 10 9.51606	27 13 9.71125	9 43
Mars	299 6	21 45 9.60013	46 22 10.02066	24 44
Jupiter	302 37	20 26 9.57119	47 13 10.03355	23 44
Saturn	348 54	6 54 9.08283	45 56 10.01423	7 11
Uranus	262 49	23 24 9.63623	3 13 8.74904	7 27
Neptune	267 47	22 14 9.61148	3 48 8.82147	9 20

PRIMARY DIRECTIONS

Examples

Direct the Sun in mundo to the conjunction with Venus mundo. The pole of Venus is 27° 13', its ascensional difference under that pole is 9° 43', which added to its R.A., 276° 6' (as Venus' declination is S.), gives its oblique ascension under its own pole =285° 49'.

The oblique ascension of Sun under the same pole is—

Pole, log. tang.	27° 13'	9.71125
Decl., log. tang.	15° 13'	9.43458
Asc. diff. log. sine	8° 2'	9.14583
R.A. of Sun	321° 12'	
O.A. of Sun	329° 14' under Venus' pole.	
O.A. Venus	285° 49' " " "	
Arc of direction =	43° 25' Sun conj.	
	Venus in mun.	

Note.—All the tangents being inserted in the speculum under the declinations and poles of the planets, they can be extracted as required.

Direct Uranus to the conjunction with the Moon in mundo.

The pole of the Moon is 50° 21', its ascensional difference under that pole is 35° 24', and its oblique descension 155° 41'.

The declin. of Uranus is	23° 24'	tan. 9.63623
Pole of Moon	50° 21'	tan. 0.08147
Asc. diff. under pole	31° 28'	sine 9.71770
Uranus' R.A..	262° 49'	
O.D. Uranus	231° 21' under Moon's pole.	
O.D. of Moon	155° 41' " " "	
Arc of direction =	75° 40' Uranus conj. Moon	
	in mundo.	

SUGGESTED METHOD OF TRUE DIRECTING

These directions take very much less time to calculate than to set out in writing, and with the speculum at hand they are readily figured out in a minute or two.

Direct the Moon to opposition of Venus in mundo.

Oblique descension of the opposition of Venus =105° 49'.

This is taken under the pole of Venus, from Venus' oblique ascension less 180° = oblique ascension of the opposite point.

Oblique descension of Moon under Venus'
    pole   134° 37'
Oblique descension of Venus under same
    pole   105° 49'
    Arc of Direction, Moon oppos. Venus
    mundo   28° 48'

These examples will doubtless serve for all conjunctions in mundo. For zodiacal directions it will be necessary to find the declination of the degree of the ecliptic held by a planet to which direction is made, or of its aspect, and add the log. tang. of this declination to the log. tang. of its pole. This will give the ascensional difference under that pole. Apply this to the right ascension to get its oblique ascension or oblique descension under that pole. The difference between this and the oblique ascension (or descension) of the planet directed, taken under the same pole, will be the arc of direction.

Planets having the same pole are either in mundane conjunction or in mundane parallel. This gives us the hint as to the calculation of mundane parallels.

Find the oblique ascension or oblique descension of the planet on which the parallel is formed, taken under its own pole. Find the oblique ascension or oblique descension (as the case may require) of the planet forming the parallel, under the same pole. The difference will be the arc of direction.

PRIMARY DIRECTIONS

Example 1. — Bring Saturn in the example horoscope to the mundane parallel of the Moon.

This direction is formed by Saturn coming up to the pole of the Moon on the other side of the meridian.

Right ascension of the Midheaven	249°	56'
Oblique descension of Moon under its own pole	155°	45'
Moon's distance from Midheaven, westward	94°	11'
Added to R.A. of M.C.	249°	56'
Oblique ascension of the parallel, eastward	344°	7'
Oblique ascension of Saturn under Moon's pole	357°	18'
Arc of direction = difference	13°	11'

This arc of direction, when computed by the semiarc method, is seen to be 8' short of the actual figures, which throws the time out nearly two months. The arc by that method is 13° 3' as compared with 13° 11', the true arc.

Example 2.—Bring Uranus to the mundane parallel of Sun in mundo. Here the planet descends the western horizon until it comes to the same pole westward as the Sun holds eastward.

Oblique ascension of the Sun under its own pole, 51° 13'	340°	59'
Subtract 180	180°	0'
Oblique descension of aspect below west horizon	160°	59'
Oblique descension of Uranus under pole of Sun	230°	14'
Arc of direction, Uranus parallel Sun mundo	69°	15'

This arc of direction by the semiarc method is found to be 70° 57', which shows an error of 1° 42', equal to one year and eight months of time.

SUGGESTED METHOD OF TRUE DIRECTING
Time Measure for Arcs

This remark brings me back again to the question of the equation of time, so much in dispute among astrologers. I think there can be little doubt that the true method is " a day for a year," which is certainly the most ancient method, as it is also the most uniform. In twenty-four hours the Earth revolves on its axis and the Sun comes again to the same meridian, having in the interval increased its longitude by more or less than a degree according to its apparent place in its orbit, *i.e.* the season of the year. The mean rate of its motion is 59' 8". Then, as all our calculations are made in terms of equatorial degrees, we have to make a proportion 59' 8" to 60', and this gives 24h. 21m. =1 year 5.334 days =1 year 5d. 8h. for each 1° in the arc of direction. Thus every 6° in the arc of direction will give an extra month, to be added to the time at the rate of 1°=1 year, which is the measure of time used in the semiarc method. If we add 5' for every 6° of arc it will come to the same thing approximately. The measure of a degree of R.A. for a year is due to Placidus. That of the Sun's mean motion, or 1° R.A. =1 year 5 days, is due to Valentine Naibod. Both are a compromise with facts. The probability is that we ought to take the measure according to the season of the year in which the birth takes place, and hence the Sun's actual increase of R.A. on that date, since the Sun is in every natural sense the great chronocrater, or time-maker. Thus, in the case of Ruskin, who was born on the 8th February, the Sun's diurnal increase of R.A. is 3' 57" = 59' 15" in arc, but its increase in longitude is 60' 43", and this being an excess 1' 35" over the mean motion in the zodiac, an arc of direction, at the rate of a day for a year, would measure to so much less, at the rate of about 11 minutes for every complete degree of the arc. It will thus be seen that the question of the validity of one method over another in primary directions does not rest entirely on the astronomical facts, but also upon the value we attach to the arcs of direction when obtained. As to the astronomy of the case, there is not the slightest doubt in my mind that the method of directing under the pole of the significator is the correct mathematical scheme. But as to the measure of time from arcs thus derived, this is a matter of experiment, and one needs to exhaust all the evidence before coming to a conclusion.

CHAPTER XVII

CONCLUSION

In the foregoing pages I have endeavoured to set out and critically examine the methods of directing advocated by Ptolemy and Placidus as modernly represented ; and I have further sought to establish their validity on general principles. I have not been blind to their imperfections, and have clearly indicated my view of the semiarc method, derived from the principles laid down by these great pioneers of a scientific astrology, when I speak of them as valuable approximations. The discrepancies are those due to incorrect use of words in describing the facts. The term " corresponding to " should be more frequently used in the semiarc method in place of the term " equal to." It is admitted that in both systems—that of proportional semiarcs and that of direction under poles—we are concerned with the apparent places of the planets in the prime vertical, and therefore when we speak of planets as being directed to a conjunction we mean an apparent conjunction as seen from the place of birth, and not either in the zodiac or by right ascension, but solely in the prime vertical or circle of observation, which coincides neither with the Equator nor the Ecliptic. Therefore, when we come to the test we find without doubt that the only way of doing this is to bring the directed body along its own arc or parallel of declination to the same pole as the promittor or body directed to. Also, it is apparent that as polar elevation is measured from the zenith in the plane of the prime vertical, planets having the same pole must be in mundane conjunction if on the same side of the meridian, or in mundane parallel if on opposite sides, which fact renders the calculation of mundane parallels a process of such extreme simplicity that I wonder it has never been pointed out before.

CONCLUSION

To correct the errors arising out of the methods of Ptolemy and Placidus, I have made a complete statement of the true doctrine of polar directions in the plane of the prime vertical, and have supplemented this by a speculum drawn according to the principles laid down, so that by mere inspection of the same, and very little figuring, all directions in mundo can be calculated. For directions in the zodiac it will be necessary to have the pole of the aspect or position in the zodiac, which can be determined by the longitudinal distance from the cusp of the house taken in proportion to the degrees of the ecliptic included in that house from the Table of Poles of the Houses, and from this we get its oblique ascension or oblique descension under its own pole and direct to it as in mundane direction.

In effect, it will be found that with a set of tables of oblique ascension, and one of tables of poles, all directions can be correctly calculated in a fraction of the time usually devoted to them, even by the very facile but faulty method of proportion of semiarcs. I have fairly stated both cases, and criticised only where criticism was necessary to correct error. In this I have done no hurt to the cause of scientific astrology, and I conclude this treatise in the earnest belief that I have even done some small service.

TABLES FOR THE USE OF ASTROLOGICAL STUDENTS

INCLUDING TABLES OF LOGARITHMIC
SINES, TANGENTS, ETC., TABLES OF
RIGHT ASCENSION, DECLINATION,
AND ASCENSIONAL DIFFERENCE,
AND TERNARY
PROPORTIONAL
LOGARITHMS

TABLES OF LOGARITHMIC
SINES, TANGENTS, ETC.

[0 degrees.] [89 degrees.]

'	Cosine.	Cotang.	Diff.	Tangent.	Diff.	Sine.	'
30	9·99998	12·05914	1424	7·94086	1424	7·94084	30
31	9·99998	12·04490	1379	7·95510	1379	7·95508	29
32	9·99998	12·03111	1336	7·96889	1336	7·96887	28
33	9·99998	12·01775	1297	7·98225	1297	7·98223	27
34	9·99998	12·00478	1259	7·99522	1259	7·99520	26
35	9·99998	11·99219	1223	8·00781	1223	8·00779	25
36	9·99998	11·97996	1190	8·02004	1190	8·02002	24
37	9·99997	11·96806	1159	8·03194	1158	8·03192	23
38	9·99997	11·95647	1128	8·04353	1128	8·04350	22
39	9·99997	11·94519	1100	8·05481	1100	8·05478	21
40	9·99997	11·93419	1072	8·06581	1072	8·06578	20
41	9·99997	11·92347	1047	8·07653	1046	8·07650	19
42	9·99997	11·91300	1022	8·08700	1022	8·08696	18
43	9·99997	11·90278	998	8·09722	999	8·09718	17
44	9·99997	11·89280	976	8·10720	976	8·10717	16
45	9·99996	11·88304	955	8·11696	954	8·11693	15
46	9·99996	11·87349	934	8·12651	934	8·12647	14
47	9·99996	11·86415	915	8·13585	914	8·13581	13
48	9·99996	11·85500	895	8·14500	896	8·14495	12
49	9·99996	11·84605	878	8·15395	877	8·15391	11
50	9·99995	11·83727	860	8·16273	860	8·16268	10
51	9·99995	11·82867	843	8·17133	843	8·17128	9
52	9·99995	11·82024	828	8·17976	827	8·17971	8
53	9·99995	11·81196	812	8·18804	812	8·18798	7
54	9·99995	11·80384	797	8·19616	797	8·19610	6
55	9·99994	11·79587	782	8·20413	782	8·20407	5
56	9·99994	11·78805	769	8·21195	769	8·21189	4
57	9·99994	11·78036	756	8·21964	755	8·21958	3
58	9·99994	11·77280	742	8·22720	743	8·22713	2
59	9·99994	11·76538	730	8·23462	730	8·23456	1
60	9·99993	11·75808		8·24192		8·24186	0
'	Sine.	Tangent.		Cotang.		Cosine.	'

101

[0 degrees.] [89 degrees.]

'	Sine.	Diff.	Tangent.	Diff.	Cotang.	Cosine.	'
0	-∞	+∞	-∞	∞	+∞	0·00000	60
1	6·46373	30103	6·46373	30103	13·53627	0·00000	59
2	6·76476	17609	6·76476	17609	13·23524	0·00000	58
3	6·94085	12494	6·94085	12494	13·05915	0·00000	57
4	7·06579	9691	7·06579	9691	12·93421	0·00000	56
5	7·16270	7918	7·16270	7918	12·83730	0·00000	55
6	7·24188	6694	7·24188	6694	12·75812	0·00000	54
7	7·30882	5800	7·30882	5800	12·69118	0·00000	53
8	7·36682	5115	7·36682	5115	12·63318	0·00000	52
9	7·41797	4576	7·41797	4576	12·58203	0·00000	51
10	7·46373	4139	7·46373	4139	12·53627	0·00000	50
11	7·50512	3779	7·50512	3779	12·49488	0·00000	49
12	7·54291	3476	7·54291	3476	12·45709	0·00000	48
13	7·57767	3219	7·57767	3219	12·42233	0·00000	47
14	7·60985	2996	7·60986	2996	12·39014	0·00000	46
15	7·63982	2802	7·63982	2803	12·36018	0·00000	45
16	7·66784	2633	7·66785	2633	12·33215	0·00000	44
17	7·69417	2483	7·69418	2482	12·30582	0·00000	43
18	7·71900	2348	7·71900	2348	12·28100	0·00000	42
19	7·74248	2227	7·74448	2228	12·25752	9·99999	41
20	7·76475	2119	7·76476	2119	12·23524	9·99999	40
21	7·78594	2021	7·78595	2020	12·21405	9·99999	39
22	7·80615	1930	7·80615	1931	12·19385	9·99999	38
23	7·82545	1848	7·82546	1848	12·17454	9·99999	37
24	7·84393	1773	7·84394	1773	12·15606	9·99999	36
25	7·86166	1704	7·86167	1704	12·13833	9·99999	35
26	7·87870	1639	7·87871	1639	12·12129	9·99999	34
27	7·89509	1579	7·89510	1579	12·10490	9·99999	33
28	7·91088	1524	7·91089	1524	12·08911	9·99999	32
29	7·92612	1472	7·92613	1473	12·07387	9·99998	31
30	7·94084		7·94086		12·05914	9·99998	30
'	Cosine.		Cotang.		Tangent.	Sine.	'

[1 degree.]   [88 degrees.]

,	Cosine.	Cotang.	Diff.	Tangent.	Diff.	Sine.	,
30	9·99985	11·58193	480	8·41807	480	8·41792	30
29	9·99985	11·57713	475	8·42287	480	8·42272	31
28	9·99984	11·57238	470	8·42762	474	8·42746	32
27	9·99984	11·56768	464	8·43232	464	8·43216	33
26	9·99984	11·56304	460	8·43696	464	8·43680	34
25	9·99983	11·55844	455	8·44156	459	8·44139	35
24	9·99983	11·55389	450	8·44611	455	8·44594	36
23	9·99983	11·54939	446	8·45061	450	8·45044	37
22	9·99982	11·54493	441	8·45507	445	8·45489	38
21	9·99982	11·54052	437	8·45948	441	8·45930	39
20	9·99982	11·53615	432	8·46385	436	8·46366	40
19	9·99981	11·53183	428	8·46817	433	8·46799	41
18	9·99981	11·52755	424	8·47245	427	8·47226	42
17	9·99981	11·52331	420	8·47669	424	8·47650	43
16	9·99980	11·51911	416	8·48089	419	8·48069	44
15	9·99980	11·51495	412	8·48505	416	8·48485	45
14	9·99979	11·51083	408	8·48917	411	8·48896	46
13	9·99979	11·50675	404	8·49325	408	8·49304	47
12	9·99979	11·50271	401	8·49729	404	8·49708	48
11	9·99978	11·49870	397	8·50130	400	8·50108	49
10	9·99978	11·49473	393	8·50527	396	8·50504	50
9	9·99977	11·49080	390	8·50920	393	8·50897	51
8	9·99977	11·48690	386	8·51310	390	8·51287	52
7	9·99977	11·48304	383	8·51696	386	8·51673	53
6	9·99976	11·47921	380	8·52079	382	8·52055	54
5	9·99976	11·47541	376	8·52459	379	8·52434	55
4	9·99975	11·47165	373	8·52835	376	8·52810	56
3	9·99975	11·46792	370	8·53208	373	8·53183	57
2	9·99974	11·46422	367	8·53578	369	8·53552	58
1	9·99974	11·46055	363	8·53945	367	8·53919	59
0	9·99974	11·45692		8·54308	363	8·54282	60
,	Sine.	Tangent.		Cotang.		Cosine.	,

[1 degree.]   [88 degrees.]

,	Cosine.	Cotang.	Diff.	Tangent.	Diff.	Sine.	,
0	9·99993	11·75808	718	8·24192	717	8·24186	60
1	9·99993	11·75090	706	8·24910	706	8·24903	59
2	9·99993	11·74384	696	8·25616	695	8·25609	58
3	9·99993	11·73688	684	8·26312	684	8·26304	57
4	9·99992	11·73004	673	8·26996	673	8·26988	56
5	9·99992	11·72331	663	8·27669	663	8·27661	55
6	9·99992	11·71668	654	8·28332	653	8·28324	54
7	9·99992	11·71014	643	8·28986	644	8·28977	53
8	9·99992	11·70371	634	8·29629	634	8·29611	52
9	9·99991	11·69737	625	8·30263	624	8·30255	51
10	9·99991	11·69112	617	8·30888	616	8·30879	50
11	9·99991	11·68495	607	8·31505	608	8·31495	49
12	9·99990	11·67888	599	8·32111	599	8·32103	48
13	9·99990	11·67289	591	8·32711	590	8·32702	47
14	9·99990	11·66698	584	8·33302	583	8·33292	46
15	9·99990	11·66114	575	8·33886	575	8·33875	45
16	9·99989	11·65539	568	8·34461	568	8·34450	44
17	9·99989	11·64971	561	8·35029	560	8·35018	43
18	9·99989	11·64410	553	8·35590	553	8·35578	42
19	9·99989	11·63857	546	8·36143	547	8·36131	41
20	9·99988	11·63311	540	8·36689	539	8·36678	40
21	9·99988	11·62771	533	8·37229	533	8·37217	39
22	9·99988	11·62238	527	8·37762	526	8·37750	38
23	9·99987	11·61711	520	8·38289	520	8·38276	37
24	9·99987	11·61191	514	8·38809	514	8·38796	36
25	9·99987	11·60677	509	8·39323	508	8·39310	35
26	9·99986	11·60168	502	8·39832	502	8·39818	34
27	9·99986	11·59666	496	8·40334	496	8·40320	33
28	9·99986	11·59170	491	8·40830	491	8·40816	32
29	9·99985	11·58679	486	8·41321	485	8·41307	31
30	9·99985	11·58193		8·41807		8·41792	30
,	Sine.	Tangent.		Cotang.		Cosine.	,

[2 degrees:]

′	Sine.	Diff.	Tangent.	Diff.	Cotang.	Cosine.	′
0	8·54282	360	8·54308	361	11·45692	9·99974	60
1	8·54642	357	8·54669	358	11·45331	9·99973	59
2	8·54999	355	8·55027	355	11·44973	9·99973	58
3	8·55354	351	8·55382	352	11·44618	9·99972	57
4	8·55705	349	8·55734	349	11·44266	9·99972	56
5	8·56054	346	8·56083	346	11·43917	9·99971	55
6	8·56400	343	8·56429	344	11·43571	9·99971	54
7	8·56743	341	8·56773	341	11·43227	9·99970	53
8	8·57084	337	8·57114	338	11·42886	9·99970	52
9	8·57421	336	8·57452	336	11·42548	9·99969	51
10	8·57757	332	8·57788	333	11·42212	9·99969	50
11	8·58089	330	8·58121	330	11·41879	9·99968	49
12	8·58419	328	8·58451	328	11·41549	9·99968	48
13	8·58747	325	8·58779	326	11·41221	9·99967	47
14	8·59072	323	8·59105	323	11·40895	9·99967	46
15	8·59395	320	8·59428	321	11·40572	9·99967	45
16	8·59715	318	8·59749	319	11·40251	9·99966	44
17	8·60033	316	8·60068	316	11·39932	9·99966	43
18	8·60349	313	8·60384	314	11·39616	9·99965	42
19	8·60662	311	8·60698	311	11·39302	9·99964	41
20	8·60973	309	8·61009	310	11·38991	9·99964	40
21	8·61282	307	8·61319	307	11·38681	9·99963	39
22	8·61589	305	8·61626	305	11·38374	9·99963	38
23	8·61894	302	8·61931	303	11·38069	9·99962	37
24	8·62196	301	8·62234	301	11·37766	9·99962	36
25	8·62497	298	8·62535	299	11·37465	9·99961	35
26	8·62795	296	8·62834	297	11·37166	9·99961	34
27	8·63091	294	8·63131	295	11·36869	9·99960	33
28	8·63385	293	8·63426	292	11·36574	9·99960	32
29	8·63678	290	8·63718	291	11·36282	9·99959	31
30	8·63968		8·64009		11·35991	9·99959	30
′	Cosine.		Cotang.		Tangent.	Sine.	′

[87 degrees.]

[2 degrees.]

′	Sine.	Diff.	Tangent.	Diff.	Cotang.	Cosine.	′
30	8·63968	288	8·64009	289	11·35991	9·99959	30
31	8·64256	287	8·64298	287	11·35702	9·99958	29
32	8·64543	284	8·64585	285	11·35415	9·99958	28
33	8·64827	283	8·64870	284	11·35130	9·99957	27
34	8·65110	281	8·65154	281	11·34846	9·99956	26
35	8·65391	279	8·65435	280	11·34565	9·99956	25
36	8·65670	277	8·65715	278	11·34285	9·99955	24
37	8·65947	276	8·65993	276	11·34007	9·99955	23
38	8·66223	274	8·66269	274	11·33731	9·99954	22
39	8·66497	272	8·66543	273	11·33457	9·99954	21
40	8·66769	270	8·66816	271	11·33184	9·99953	20
41	8·67039	269	8·67087	269	11·32913	9·99952	19
42	8·67308	267	8·67356	268	11·32644	9·99952	18
43	8·67575	266	8·67624	266	11·32376	9·99951	17
44	8·67841	263	8·67890	264	11·32110	9·99951	16
45	8·68104	263	8·68154	263	11·31846	9·99950	15
46	8·68367	260	8·68417	261	11·31583	9·99949	14
47	8·68627	259	8·68678	260	11·31322	9·99949	13
48	8·68886	258	8·68938	258	11·31062	9·99948	12
49	8·69144	256	8·69196	257	11·30804	9·99948	11
50	8·69400	254	8·69453	255	11·30547	9·99947	10
51	8·69654	253	8·69708	254	11·30292	9·99946	9
52	8·69907	252	8·69962	252	11·30038	9·99946	8
53	8·70159	250	8·70214	251	11·29786	9·99945	7
54	8·70409	249	8·70465	249	11·29535	9·99944	6
55	8·70658	247	8·70714	248	11·29286	9·99944	5
56	8·70905	246	8·70962	246	11·29038	9·99943	4
57	8·71151	244	8·71208	245	11·28792	9·99942	3
58	8·71395	243	8·71453	244	11·28547	9·99942	2
59	8·71638	242	8·71697	243	11·28303	9·99941	1
60	8·71880		8·71940		11·28060	9·99940	0
′	Cosine.		Cotang.		Tangent.	Sine.	′

[87 degrees.]

[3 degrees.]   **[86 degrees.]**

′	Cosine.	Cotang.	Diff.	Tangent.	Diff.	Sine.	′
30	9·99919	11·21351	206	8·78649	206	8·78568	30
31	9·99918	11·21145	206	8·78855	205	8·78774	29
32	9·99917	11·20939	206	8·79061	205	8·78979	28
33	9·99917	11·20734	205	8·79266	204	8·79183	27
34	9·99916	11·20530	204	8·79470	203	8·79386	26
35	9·99915	11·20327	203	8·79673	202	8·79588	25
36	9·99914	11·20125	202	8·79875	201	8·79789	24
37	9·99913	11·19924	201	8·80076	201	8·79990	23
38	9·99913	11·19723	201	8·80277	199	8·80189	22
39	9·99912	11·19524	199	8·80476	199	8·80388	21
40	9·99911	11·19326	198	8·80674	197	8·80585	20
41	9·99910	11·19128	198	8·80872	197	8·80782	19
42	9·99909	11·18932	196	8·81068	196	8·80978	18
43	9·99909	11·18736	196	8·81264	195	8·81173	17
44	9·99908	11·18541	195	8·81459	194	8·81367	16
45	9·99907	11·18347	194	8·81653	193	8·81560	15
46	9·99906	11·18154	193	8·81846	192	8·81752	14
47	9·99905	11·17962	192	8·82038	192	8·81944	13
48	9·99904	11·17770	192	8·82230	190	8·82134	12
49	9·99904	11·17580	190	8·82420	190	8·82324	11
50	9·99903	11·17390	190	8·82610	189	8·82513	10
51	9·99902	11·17201	189	8·82799	188	8·82701	9
52	9·99901	11·17013	188	8·82987	187	8·82888	8
53	9·99900	11·16825	188	8·83175	186	8·83075	7
54	9·99899	11·16639	186	8·83361	185	8·83261	6
55	9·99898	11·16453	186	8·83547	184	8·83446	5
56	9·99898	11·16268	185	8·83732	184	8·83630	4
57	9·99897	11·16084	184	8·83916	183	8·83813	3
58	9·99896	11·15900	184	8·84100	183	8·83996	2
59	9·99895	11·15718	182	8·84282	181	8·84177	1
60	9·99894	11·15536	182	8·84464	181	8·84358	0
′	Sine.	Tangent.		Cotang.		Cosine.	′

[3 degrees.]   **[86 degrees.]**

′	Cosine.	Cotang.	Diff.	Tangent.	Diff.	Sine.	′
0	9·99940	11·28060	241	8·71940	240	8·71880	60
1	9·99940	11·27819	239	8·72181	239	8·72120	59
2	9·99939	11·27580	239	8·72420	238	8·72359	58
3	9·99938	11·27341	237	8·72659	237	8·72597	57
4	9·99938	11·27104	236	8·72896	235	8·72834	56
5	9·99937	11·26868	234	8·73132	234	8·73069	55
6	9·99936	11·26634	234	8·73366	232	8·73303	54
7	9·99936	11·26400	232	8·73600	232	8·73535	53
8	9·99935	11·26168	231	8·73832	230	8·73767	52
9	9·99934	11·25937	229	8·74063	229	8·73997	51
10	9·99934	11·25708	229	8·74292	228	8·74226	50
11	9·99933	11·25479	227	8·74521	226	8·74454	49
12	9·99932	11·25252	226	8·74748	226	8·74680	48
13	9·99932	11·25026	225	8·74974	224	8·74906	47
14	9·99931	11·24801	224	8·75199	223	8·75130	46
15	9·99930	11·24577	222	8·75423	222	8·75353	45
16	9·99929	11·24355	222	8·75645	220	8·75575	44
17	9·99929	11·24133	220	8·75867	220	8·75795	43
18	9·99928	11·23913	219	8·76087	219	8·76015	42
19	9·99927	11·23694	219	8·76306	217	8·76234	41
20	9·99926	11·23475	217	8·76525	216	8·76451	40
21	9·99926	11·23258	216	8·76742	216	8·76667	39
22	9·99925	11·23042	215	8·76958	214	8·76883	38
23	9·99924	11·22827	214	8·77173	213	8·77097	37
24	9·99923	11·22613	213	8·77387	212	8·77310	36
25	9·99923	11·22400	211	8·77600	211	8·77522	35
26	9·99922	11·22189	211	8·77811	210	8·77733	34
27	9·99921	11·21978	210	8·78022	209	8·77943	33
28	9·99920	11·21768	209	8·78232	208	8·78152	32
29	9·99920	11·21559	208	8·78441	208	8·78360	31
30	9·99919	11·21351		8·78649		8·78568	30
′	Sine.	Tangent.		Cotang.		Cosine.	′

[4 degrees.]

'	Sine.	Diff.	Tangent.	Diff.	Cotang.	Cosine.	'
30	8·89464	161	8·89598	162	11·10402	9·99866	30
31	8·89625	159	8·89760	160	11·10240	9·99865	29
32	8·89784	159	8·89920	160	11·10080	9·99864	28
33	8·89943	159	8·90080	160	11·09920	9·99863	27
34	8·90102	158	8·90240	159	11·09760	9·99862	26
35	8·90260	157	8·90399	158	11·09601	9·99861	25
36	8·90417	157	8·90557	158	11·09443	9·99860	24
37	8·90574	156	8·90715	157	11·09285	9·99859	23
38	8·90730	155	8·90872	157	11·09128	9·99858	22
39	8·90885	155	8·91029	156	11·08971	9·99857	21
40	8·91040	155	8·91185	155	11·08815	9·99856	20
41	8·91195	154	8·91340	155	11·08660	9·99855	19
42	8·91349	153	8·91495	155	11·08505	9·99854	18
43	8·91502	153	8·91650	153	11·08350	9·99853	17
44	8·91655	152	8·91803	153	11·08197	9·99852	16
45	8·91807	152	8·91957	152	11·08043	9·99851	15
46	8·91959	151	8·92110	152	11·07890	9·99850	14
47	8·92110	151	8·92262	151	11·07738	9·99848	13
48	8·92261	150	8·92414	151	11·07586	9·99847	12
49	8·92411	150	8·92565	151	11·07435	9·99846	11
50	8·92561	149	8·92716	150	11·07284	9·99845	10
51	8·92710	149	8·92866	150	11·07134	9·99844	9
52	8·92859	148	8·93016	149	11·06984	9·99843	8
53	8·93007	147	8·93165	148	11·06835	9·99842	7
54	8·93154	147	8·93313	149	11·06687	9·99841	6
55	8·93301	147	8·93462	147	11·06538	9·99840	5
56	8·93448	146	8·93609	147	11·06391	9·99839	4
57	8·93594	146	8·93756	147	11·06244	9·99838	3
58	8·93740	145	8·93903	146	11·06097	9·99837	2
59	8·93885	145	8·94049	146	11·05951	9·99836	1
60	8·94030		8·94195		11·05805	9·99834	0
'	Cosine.		Cotang.		Tangent.	Sine.	'

[85 degrees.]

105

[4 degrees.]

'	Sine.	Diff.	Tangent.	Diff.	Cotang.	Cosine.	'
0	8·84358	181	8·84464	182	11·15536	9·99894	60
1	8·84539	179	8·84646	180	11·15354	9·99893	59
2	8·84718	179	8·84826	180	11·15174	9·99892	58
3	8·84897	178	8·85006	179	11·14994	9·99891	57
4	8·85075	177	8·85185	178	11·14815	9·99891	56
5	8·85252	177	8·85363	177	11·14637	9·99890	55
6	8·85429	176	8·85540	177	11·14460	9·99889	54
7	8·85605	175	8·85717	176	11·14283	9·99888	53
8	8·85780	175	8·85893	176	11·14107	9·99887	52
9	8·85955	173	8·86069	174	11·13931	9·99886	51
10	8·86128	173	8·86243	174	11·13757	9·99885	50
11	8·86301	173	8·86417	174	11·13583	9·99884	49
12	8·86474	171	8·86591	172	11·13409	9·99883	48
13	8·86645	171	8·86763	172	11·13237	9·99882	47
14	8·86816	171	8·86935	171	11·13065	9·99881	46
15	8·86987	169	8·87106	171	11·12894	9·99880	45
16	8·87156	169	8·87277	170	11·12723	9·99879	44
17	8·87325	169	8·87447	169	11·12553	9·99878	43
18	8·87494	167	8·87616	169	11·12384	9·99878	42
19	8·87661	168	8·87785	168	11·12215	9·99877	41
20	8·87829	166	8·87953	167	11·12047	9·99876	40
21	8·87995	166	8·88120	167	11·11880	9·99875	39
22	8·88161	165	8·88287	166	11·11713	9·99874	38
23	8·88326	164	8·88453	165	11·11547	9·99873	37
24	8·88490	164	8·88618	165	11·11382	9·99872	36
25	8·88654	163	8·88783	163	11·11217	9·99871	35
26	8·88817	163	8·88948	163	11·11052	9·99870	34
27	8·88980	162	8·89111	163	11·10889	9·99869	33
28	8·89142	162	8·89274	163	11·10726	9·99868	32
29	8·89304	160	8·89437	161	11·10563	9·99867	31
30	8·89464		8·89598		11·10402	9·99866	30
'	Cosine.		Cotang.		Tangent.	Sine.	'

[85 degrees.]

[5 degrees.]   [84 degrees.]

′	Sine.	Diff.	Tangent.	Diff.	Cotang.	Cosine.	′
30	8·98157	131	8·98358	132	11·01642	9·99800	30
31	8·98288	131	8·98490	132	11·01510	9·99798	29
32	8·98419	130	8·98622	131	11·01378	9·99797	28
33	8·98549	130	8·98753	131	11·01247	9·99796	27
34	8·98679	129	8·98884	131	11·01116	9·99795	26
35	8·98808	129	8·99015	130	11·00985	9·99793	25
36	8·98937	129	8·99145	130	11·00855	9·99792	24
37	8·99066	128	8·99275	130	11·00725	9·99791	23
38	8·99194	128	8·99405	129	11·00595	9·99790	22
39	8·99322	128	8·99534	128	11·00466	9·99788	21
40	8·99450	127	8·99662	129	11·00338	9·99787	20
41	8·99577	127	8·99791	128	11·00209	9·99786	19
42	8·99704	126	8·99919	127	11·00081	9·99785	18
43	8·99830	126	9·00046	128	10·99954	9·99783	17
44	8·99956	126	9·00174	127	10·99826	9·99782	16
45	9·00082	125	9·00301	126	10·99699	9·99781	15
46	9·00207	125	9·00427	126	10·99573	9·99780	14
47	9·00332	124	9·00553	126	10·99447	9·99778	13
48	9·00456	125	9·00679	126	10·99321	9·99777	12
49	9·00581	123	9·00805	125	10·99195	9·99776	11
50	9·00704	124	9·00930	125	10·99070	9·99775	10
51	9·00828	123	9·01055	124	10·98945	9·99773	9
52	9·00951	123	9·01179	124	10·98821	9·99772	8
53	9·01074	122	9·01303	124	10·98697	9·99771	7
54	9·01196	122	9·01427	123	10·98573	9·99769	6
55	9·01318	122	9·01550	123	10·98450	9·99768	5
56	9·01440	121	9·01673	123	10·98327	9·99767	4
57	9·01561	121	9·01796	122	10·98204	9·99765	3
58	9·01682	121	9·01918	122	10·98082	9·99764	2
59	9·01803	120	9·02040	122	10·97960	9·99763	1
60	9·01923		9·02162		10·97838	9·99761	0
′	Cosine.		Cotang.		Tangent.	Sine.	′

[84 degrees.]

106

[5 degrees.]   [84 degrees.]

′	Sine.	Diff.	Tangent.	Diff.	Cotang.	Cosine.	′
0	8·94030	144	8·94195	145	11·05805	9·99834	60
1	8·94174	143	8·94340	145	11·05660	9·99833	59
2	8·94317	144	8·94485	145	11·05515	9·99832	58
3	8·94461	142	8·94630	143	11·05370	9·99831	57
4	8·94603	143	8·94773	144	11·05227	9·99830	56
5	8·94746	141	8·94917	143	11·05083	9·99829	55
6	8·94887	142	8·95060	142	11·04940	9·99828	54
7	8·95029	141	8·95202	142	11·04798	9·99827	53
8	8·95170	140	8·95344	142	11·04656	9·99825	52
9	8·95310	140	8·95486	141	11·04514	9·99824	51
10	8·95450	139	8·95627	140	11·04373	9·99823	50
11	8·95589	139	8·95767	141	11·04233	9·99822	49
12	8·95728	139	8·95908	139	11·04092	9·99821	48
13	8·95867	138	8·96047	139	11·03953	9·99820	47
14	8·96005	138	8·96187	138	11·03813	9·99819	46
15	8·96143	137	8·96325	139	11·03675	9·99817	45
16	8·96280	137	8·96464	138	11·03536	9·99816	44
17	8·96417	136	8·96602	137	11·03398	9·99815	43
18	8·96553	136	8·96739	138	11·03261	9·99814	42
19	8·96689	136	8·96877	136	11·03123	9·99813	41
20	8·96825	135	8·97013	137	11·02987	9·99812	40
21	8·96960	135	8·97150	135	11·02850	9·99810	39
22	8·97095	134	8·97285	136	11·02715	9·99809	38
23	8·97229	134	8·97421	135	11·02579	9·99808	37
24	8·97363	133	8·97556	135	11·02444	9·99807	36
25	8·97496	133	8·97691	134	11·02309	9·99806	35
26	8·97629	133	8·97825	134	11·02175	9·99804	34
27	8·97762	132	8·97959	133	11·02041	9·99803	33
28	8·97894	132	8·98092	133	11·01908	9·99802	32
29	8·98026	131	8·98225	133	11·01775	9·99801	31
30	8·98157		8·98358		11·01642	9·99800	30
′	Cosine.		Cotang.		Tangent.	Sine.	′

[84 degrees.]

[6 degrees.] [83 degrees.]

′	Sine	Diff.	Tangent	Diff.	Cotang.	Cosine.	′
30	9·05386	111	9·05666	112	10·94334	9·99720	30
31	9·05497	110	9·05778	112	10·94222	9·99718	29
32	9·05607	110	9·05890	112	10·94110	9·99717	28
33	9·05717	110	9·06002	111	10·93998	9·99716	27
34	9·05827	110	9·06113	111	10·93887	9·99714	26
35	9·05937	109	9·06224	111	10·93776	9·99713	25
36	9·06046	109	9·06335	110	10·93665	9·99711	24
37	9·06155	109	9·06445	111	10·93555	9·99710	23
38	9·06264	108	9·06556	110	10·93444	9·99708	22
39	9·06372	109	9·06666	109	10·93334	9·99707	21
40	9·06481	108	9·06775	110	10·93225	9·99705	20
41	9·06589	107	9·06885	109	10·93115	9·99704	19
42	9·06696	108	9·06994	109	10·93006	9·99702	18
43	9·06804	107	9·07103	108	10·92897	9·99701	17
44	9·06911	107	9·07211	109	10·92789	9·99699	16
45	9·07018	106	9·07320	108	10·92680	9·99698	15
46	9·07124	107	9·07428	108	10·92572	9·99696	14
47	9·07231	106	9·07536	107	10·92464	9·99695	13
48	9·07337	105	9·07643	108	10·92357	9·99693	12
49	9·07442	106	9·07751	107	10·92249	9·99692	11
50	9·07548	105	9·07858	106	10·92142	9·99690	10
51	9·07653	105	9·07964	107	10·92036	9·99689	9
52	9·07758	105	9·08071	106	10·91929	9·99687	8
53	9·07863	105	9·08177	106	10·91823	9·99686	7
54	9·07968	104	9·08283	106	10·91717	9·99684	6
55	9·08072	104	9·08389	106	10·91611	9·99683	5
56	9·08176	104	9·08495	105	10·91505	9·99681	4
57	9·08280	103	9·08600	105	10·91400	9·99680	3
58	9·08383	103	9·08705	105	10·91295	9·99678	2
59	9·08486	103	9·08810	104	10·91190	9·99677	1
60	9·08589		9·08914		10·91086	9·99675	0
′	Cosine.		Cotang.		Tangent.	Sine.	′

[6 degrees.] [83 degrees.]

′	Sine	Diff.	Tangent	Diff.	Cotang.	Cosine.	′
0	9·01923	120	9·02162	121	10·97838	9·99761	60
1	9·02043	120	9·02283	121	10·97717	9·99760	59
2	9·02163	120	9·02404	121	10·97596	9·99759	58
3	9·02283	119	9·02525	120	10·97475	9·99757	57
4	9·02402	118	9·02645	121	10·97355	9·99756	56
5	9·02520	119	9·02766	119	10·97234	9·99755	55
6	9·02639	118	9·02885	120	10·97115	9·99753	54
7	9·02757	117	9·03005	119	10·96995	9·99752	53
8	9·02874	118	9·03124	118	10·96876	9·99751	52
9	9·02992	117	9·03242	119	10·96758	9·99749	51
10	9·03109	117	9·03361	118	10·96639	9·99748	50
11	9·03226	116	9·03479	118	10·96521	9·99747	49
12	9·03342	116	9·03597	117	10·96403	9·99745	48
13	9·03458	116	9·03714	118	10·96286	9·99744	47
14	9·03574	116	9·03832	116	10·96168	9·99742	46
15	9·03690	115	9·03948	117	10·96052	9·99741	45
16	9·03805	115	9·04065	116	10·95935	9·99740	44
17	9·03920	114	9·04181	116	10·95819	9·99738	43
18	9·04034	115	9·04297	116	10·95703	9·99737	42
19	9·04149	113	9·04413	115	10·95587	9·99736	41
20	9·04262	114	9·04528	115	10·95472	9·99734	40
21	9·04376	114	9·04643	115	10·95357	9·99733	39
22	9·04490	113	9·04758	115	10·95242	9·99731	38
23	9·04603	112	9·04873	114	10·95127	9·99730	37
24	9·04715	113	9·04987	114	10·95013	9·99728	36
25	9·04828	112	9·05101	113	10·94899	9·99727	35
26	9·04940	112	9·05214	114	10·94786	9·99726	34
27	9·05052	112	9·05328	113	10·94672	9·99724	33
28	9·05164	111	9·05441	112	10·94559	9·99723	32
29	9·05275	111	9·05553	113	10·94447	9·99721	31
30	9·05386		9·05666		10·94334	9·99720	30
′	Cosine.		Cotang.		Tangent.	Sine.	′

[7 degrees.]

'	Sine.	Diff.	Tangent.	Diff.	Cotang.	Cosine.	D.	'
30	9·11570	96	9·11943	97	10·88057	9·99627	2	30
31	9·11666	95	9·12040	98	10·87960	9·99625	1	29
32	9·11761	96	9·12138	97	10·87862	9·99624	2	28
33	9·11857	95	9·12235	97	10·87765	9·99622	2	27
34	9·11952	95	9·12332	96	10·87668	9·99620	2	26
35	9·12047	95	9·12428	97	10·87572	9·99618	1	25
36	9·12142	94	9·12525	96	10·87475	9·99617	2	24
37	9·12236	95	9·12621	96	10·87379	9·99615	2	23
38	9·12331	94	9·12717	96	10·87283	9·99613	2	22
39	9·12425	94	9·12813	96	10·87187	9·99612	1	21
40	9·12519	93	9·12909	95	10·87091	9·99610	2	20
41	9·12612	94	9·13004	95	10·86996	9·99608	2	19
42	9·12706	93	9·13099	95	10·86901	9·99607	2	18
43	9·12799	93	9·13194	95	10·86806	9·99605	2	17
44	9·12892	93	9·13289	95	10·86711	9·99603	2	16
45	9·12985	93	9·13384	94	10·86616	9·99601	1	15
46	9·13078	93	9·13478	95	10·86522	9·99600	2	14
47	9·13171	92	9·13573	94	10·86427	9·99598	2	13
48	9·13263	92	9·13667	94	10·86333	9·99596	1	12
49	9·13355	92	9·13761	93	10·86239	9·99595	2	11
50	9·13447	92	9·13854	93	10·86146	9·99593	2	10
51	9·13539	91	9·13948	93	10·86052	9·99591	2	9
52	9·13630	92	9·14041	93	10·85959	9·99589	1	8
53	9·13722	91	9·14134	93	10·85866	9·99588	2	7
54	9·13813	91	9·14227	93	10·85773	9·99586	2	6
55	9·13904	90	9·14320	92	10·85680	9·99584	2	5
56	9·13994	91	9·14412	92	10·85588	9·99582	1	4
57	9·14085	90	9·14504	93	10·85496	9·99581	2	3
58	9·14175	91	9·14597	91	10·85403	9·99579	2	2
59	9·14266	90	9·14688	92	10·85312	9·99577	2	1
60	9·14356		9·14780		10·85220	9·99575		0
'	Cosine.		Cotang.		Tangent.	Sine.		'

[82 degrees.]

108

[7 degrees.]

'	Sine.	Diff.	Tangent.	Diff.	Cotang.	Cosine.	'
0	9·08589	103	9·08914	105	10·91086	9·99675	60
1	9·08692	103	9·09019	105	10·90981	9·99674	59
2	9·08795	102	9·09123	104	10·90877	9·99672	58
3	9·08897	102	9·09227	104	10·90773	9·99670	57
4	9·08999	102	9·09330	103	10·90670	9·99669	56
5	9·09101	101	9·09434	103	10·90566	9·99667	55
6	9·09202	102	9·09537	103	10·90463	9·99666	54
7	9·09304	101	9·09640	102	10·90360	9·99664	53
8	9·09405	101	9·09742	103	10·90258	9·99663	52
9	9·09506	100	9·09845	102	10·90155	9·99661	51
10	9·09606	101	9·09947	102	10·90053	9·99659	50
11	9·09707	100	9·10049	101	10·89951	9·99658	49
12	9·09807	100	9·10150	102	10·89850	9·99656	48
13	9·09907	99	9·10252	101	10·89748	9·99655	47
14	9·10006	100	9·10353	101	10·89647	9·99653	46
15	9·10106	99	9·10454	101	10·89546	9·99651	45
16	9·10205	99	9·10555	101	10·89445	9·99650	44
17	9·10304	98	9·10656	100	10·89344	9·99648	43
18	9·10402	99	9·10756	100	10·89244	9·99647	42
19	9·10501	98	9·10856	100	10·89144	9·99645	41
20	9·10599	98	9·10956	100	10·89044	9·99643	40
21	9·10697	98	9·11056	99	10·88944	9·99642	39
22	9·10795	98	9·11155	99	10·88845	9·99640	38
23	9·10893	97	9·11254	99	10·88746	9·99638	37
24	9·10990	97	9·11353	99	10·88647	9·99637	36
25	9·11087	97	9·11452	99	10·88548	9·99635	35
26	9·11184	97	9·11551	98	10·88449	9·99633	34
27	9·11281	96	9·11649	98	10·88351	9·99632	33
28	9·11377	97	9·11747	98	10·88253	9·99630	32
29	9·11474	96	9·11845	98	10·88155	9·99629	31
30	9·11570		9·11943		10·88057	9·99627	30
'	Cosine.		Cotang.		Tangent.	Sine.	'

[82 degrees.]

[8 degrees.] **[81 degrees.]**

'	D.	Cosine.	Cotang.	Diff.	Tangent.	Diff.	Sine.	'
30	2	9.99520	10.82550	86	9.17450	85	9.16970	30
31	1	9.99518	10.82464	86	9.17536	85	9.17055	29
32	2	9.99516	10.82378	86	9.17622	84	9.17139	28
33	2	9.99515	10.82292	86	9.17708	84	9.17223	27
34	2	9.99513	10.82206	86	9.17794	84	9.17307	26
35	2	9.99511	10.82120	85	9.17880	84	9.17391	25
36	2	9.99509	10.82035	86	9.17965	83	9.17474	24
37	2	9.99507	10.81949	85	9.18051	84	9.17558	23
38	2	9.99505	10.81864	85	9.18136	83	9.17641	22
39	2	9.99503	10.81779	85	9.18221	83	9.17724	21
40	1	9.99501	10.81694	85	9.18306	83	9.17807	20
41	2	9.99499	10.81609	84	9.18391	83	9.17890	19
42	2	9.99497	10.81525	85	9.18475	82	9.17973	18
43	2	9.99495	10.81440	84	9.18560	82	9.18055	17
44	2	9.99494	10.81356	84	9.18644	83	9.18137	16
45	2	9.99492	10.81272	84	9.18728	82	9.18220	15
46	2	9.99490	10.81188	84	9.18812	81	9.18302	14
47	2	9.99488	10.81104	83	9.18896	82	9.18383	13
48	2	9.99486	10.81021	84	9.18979	81	9.18465	12
49	2	9.99484	10.80937	83	9.19063	81	9.18547	11
50	2	9.99482	10.80854	83	9.19146	81	9.18628	10
51	2	9.99480	10.80771	83	9.19229	81	9.18709	9
52	2	9.99478	10.80688	83	9.19312	81	9.18790	8
53	2	9.99476	10.80605	83	9.19395	81	9.18871	7
54	2	9.99474	10.80522	83	9.19478	81	9.18952	6
55	2	9.99472	10.80439	82	9.19561	80	9.19033	5
56	2	9.99470	10.80357	82	9.19643	80	9.19113	4
57	2	9.99468	10.80275	82	9.19725	80	9.19193	3
58	2	9.99466	10.80193	82	9.19807	80	9.19273	2
59	2	9.99464	10.80111	82	9.19889	80	9.19353	1
60	2	9.99462	10.80029		9.19971		9.19433	0
		Sine.	Tangent.		Cotang.			'

109

[8 degrees.] **[81 degrees.]**

'	D.	Cosine.	Cotang.	Diff.	Tangent.	Diff.	Sine.	'
0	1	9.99575	10.85220	92	9.14780	89	9.14356	60
1	2	9.99574	10.85128	91	9.14872	90	9.14445	59
2	2	9.99572	10.85037	91	9.14963	90	9.14535	58
3	2	9.99570	10.84946	91	9.15054	89	9.14624	57
4	2	9.99568	10.84855	91	9.15145	90	9.14714	56
5	1	9.99566	10.84764	91	9.15236	89	9.14803	55
6	2	9.99565	10.84673	90	9.15327	88	9.14891	54
7	2	9.99563	10.84583	91	9.15417	89	9.14980	53
8	2	9.99561	10.84492	90	9.15508	89	9.15069	52
9	2	9.99559	10.84402	90	9.15598	88	9.15157	51
10	1	9.99557	10.84312	89	9.15688	88	9.15245	50
11	2	9.99556	10.84223	90	9.15777	88	9.15333	49
12	2	9.99554	10.84133	89	9.15867	88	9.15421	48
13	2	9.99552	10.84044	90	9.15956	87	9.15508	47
14	2	9.99550	10.83954	89	9.16046	88	9.15596	46
15	2	9.99548	10.83865	89	9.16135	87	9.15683	45
16	2	9.99546	10.83776	88	9.16224	87	9.15770	44
17	1	9.99545	10.83688	89	9.16312	87	9.15857	43
18	2	9.99543	10.83599	88	9.16401	87	9.15944	42
19	2	9.99541	10.83511	88	9.16489	86	9.16030	41
20	2	9.99539	10.83423	88	9.16577	86	9.16116	40
21	2	9.99537	10.83335	88	9.16665	87	9.16203	39
22	1	9.99535	10.83247	88	9.16753	86	9.16289	38
23	2	9.99533	10.83159	87	9.16841	85	9.16374	37
24	2	9.99532	10.83072	88	9.16928	86	9.16460	36
25	2	9.99530	10.82984	87	9.17016	85	9.16545	35
26	2	9.99528	10.82897	87	9.17103	86	9.16631	34
27	1	9.99526	10.82810	87	9.17190	85	9.16716	33
28	2	9.99523	10.82723	86	9.17277	85	9.16801	32
29	2	9.99522	10.82637	87	9.17363	84	9.16886	31
30	2	9.99520	10.82550		9.17450		9.16970	30
		Sine.	Tangent.		Cotang.			'

[9 degrees]   **[80 degrees]**

,	D.	Cosine.	Cotang.	Diff.	Tangent.	Diff.	Sine.	,
30	2	9.99400	10.77619	77	9.22361	75	9.21761	30
29	2	9.99398	10.77562	78	9.22438	76	9.21836	31
28	2	9.99396	10.77484	77	9.22516	75	9.21912	32
27	2	9.99394	10.77407	77	9.22593	75	9.21987	33
26	2	9.99392	10.77330	77	9.22670	75	9.22062	34
25	2	9.99390	10.77253	77	9.22747	74	9.22137	35
24	3	9.99388	10.77176	77	9.22824	75	9.22211	36
23	2	9.99385	10.77099	76	9.22901	75	9.22286	37
22	2	9.99383	10.77023	77	9.22977	74	9.22361	38
21	2	9.99381	10.76946	76	9.23054	74	9.22435	39
20	2	9.99379	10.76870	76	9.23130	74	9.22509	40
19	2	9.99377	10.76794	76	9.23206	74	9.22583	41
18	2	9.99375	10.76717	76	9.23283	74	9.22657	42
17	3	9.99372	10.76641	76	9.23359	74	9.22731	43
16	2	9.99370	10.76565	75	9.23435	73	9.22805	44
15	2	9.99368	10.76490	76	9.23510	74	9.22878	45
14	2	9.99366	10.76414	75	9.23586	73	9.22952	46
13	3	9.99364	10.76339	76	9.23661	73	9.23025	47
12	2	9.99362	10.76263	75	9.23737	73	9.23098	48
11	2	9.99359	10.76188	75	9.23812	73	9.23171	49
10	2	9.99357	10.76113	75	9.23887	73	9.23244	50
9	2	9.99355	10.76038	75	9.23962	73	9.23317	51
8	2	9.99353	10.75963	75	9.24037	72	9.23390	52
7	2	9.99351	10.75888	74	9.24112	73	9.23462	53
6	2	9.99348	10.75814	75	9.24186	72	9.23535	54
5	2	9.99346	10.75739	74	9.24261	72	9.23607	55
4	2	9.99344	10.75665	74	9.24335	72	9.23679	56
3	2	9.99342	10.75590	74	9.24410	71	9.23752	57
2	3	9.99340	10.75516	74	9.24484	72	9.23823	58
1	2	9.99337	10.75442	74	9.24558	72	9.23895	59
0	2	9.99335	10.75368		9.24632		9.23967	60
		Sine.	Tangent.		Cotang.		Cosine.	,

[9 degrees]   **[80 degrees]**

,	D.	Cosine.	Cotang.	Diff.	Tangent.	Diff.	Sine.	,
60	2	9.99462	10.80029	82	9.19971	80	9.19433	0
59	2	9.99460	10.79947	81	9.20053	79	9.19513	1
58	2	9.99458	10.79866	82	9.20134	80	9.19592	2
57	2	9.99456	10.79784	81	9.20216	79	9.19672	3
56	2	9.99454	10.79703	81	9.20297	79	9.19751	4
55	2	9.99452	10.79622	81	9.20378	79	9.19830	5
54	2	9.99450	10.79541	81	9.20459	79	9.19909	6
53	2	9.99448	10.79460	81	9.20540	79	9.19988	7
52	2	9.99446	10.79379	80	9.20621	79	9.20067	8
51	2	9.99444	10.79299	81	9.20701	78	9.20145	9
50	2	9.99442	10.79218	80	9.20782	78	9.20223	10
49	2	9.99440	10.79138	80	9.20862	80	9.20302	11
48	2	9.99438	10.79058	80	9.20942	80	9.20380	12
47	2	9.99436	10.78978	80	9.21022	80	9.20458	13
46	2	9.99434	10.78898	80	9.21102	77	9.20535	14
45	2	9.99432	10.78818	79	9.21182	78	9.20613	15
44	3	9.99429	10.78739	80	9.21261	78	9.20691	16
43	2	9.99427	10.78659	79	9.21341	77	9.20768	17
42	2	9.99425	10.78580	79	9.21420	77	9.20845	18
41	2	9.99423	10.78501	79	9.21499	77	9.20922	19
40	2	9.99421	10.78422	79	9.21578	77	9.20999	20
39	3	9.99419	10.78343	79	9.21657	77	9.21076	21
38	2	9.99417	10.78264	78	9.21736	77	9.21153	22
37	2	9.99415	10.78186	79	9.21814	79	9.21229	23
36	3	9.99413	10.78107	78	9.21893	77	9.21306	24
35	2	9.99411	10.78029	78	9.21971	76	9.21382	25
34	2	9.99409	10.77951	78	9.22049	76	9.21458	26
33	3	9.99407	10.77873	78	9.22127	76	9.21534	27
32	2	9.99404	10.77795	78	9.22205	76	9.21610	28
31	2	9.99402	10.77717	78	9.22283	76	9.21685	29
30	2	9.99400	10.77639		9.22361		9.21761	30
		Sine.	Tangent.		Cotang.		Cosine.	,

[10 degrees.] [79 degrees.]

′	Sine	Diff	Tangent	Diff	Cotang.	Cosine	D	′
30	9·26063	68	9·26797	70	10·73203	9·99267	3	30
31	9·26131	68	9·26867	70	10·73133	9·99264	3	29
32	9·26199	68	9·26937	70	10·73063	9·99262	2	28
33	9·26267	68	9·27008	71	10·72992	9·99260	3	27
34	9·26335	68	9·27078	70	10·72922	9·99257	3	26
35	9·26403	67	9·27148	70	10·72852	9·99255	2	25
36	9·26470	68	9·27218	70	10·72782	9·99252	3	24
37	9·26538	67	9·27288	69	10·72712	9·99250	2	23
38	9·26605	68	9·27357	70	10·72643	9·99248	3	22
39	9·26673	67	9·27427	69	10·72573	9·99245	2	21
40	9·26740	67	9·27496	69	10·72504	9·99243	3	20
41	9·26806	67	9·27566	69	10·72434	9·99241	3	19
42	9·26873	67	9·27635	69	10·72365	9·99238	2	18
43	9·26940	67	9·27704	69	10·72296	9·99236	3	17
44	9·27007	66	9·27773	69	10·72227	9·99233	2	16
45	9·27073	67	9·27842	69	10·72158	9·99231	3	15
46	9·27140	66	9·27911	69	10·72089	9·99229	3	14
47	9·27206	67	9·27980	69	10·72020	9·99226	2	13
48	9·27273	66	9·28049	68	10·71951	9·99224	3	12
49	9·27339	66	9·28117	69	10·71883	9·99221	2	11
50	9·27405	66	9·28186	68	10·71814	9·99219	3	10
51	9·27471	66	9·28254	69	10·71746	9·99217	3	9
52	9·27537	65	9·28323	68	10·71677	9·99214	2	8
53	9·27602	66	9·28391	68	10·71609	9·99212	3	7
54	9·27668	66	9·28459	68	10·71541	9·99209	2	6
55	9·27734	65	9·28527	68	10·71473	9·99207	3	5
56	9·27799	65	9·28595	67	10·71405	9·99204	2	4
57	9·27864	66	9·28662	68	10·71338	9·99202	3	3
58	9·27930	65	9·28730	68	10·71270	9·99200	2	2
59	9·27995	65	9·28798	67	10·71202	9·99197	3	1
60	9·28060		9·28865		10·71135	9·99195		0
′	Cosine.		Cotang.		Tangent.	Sine.	D.	′

111

[10 degrees.] [79 degrees.]

′	Sine	Diff	Tangent	Diff	Cotang.	Cosine	D.	′
0	9·23967	72	9·24632	74	10·75368	9·99335	2	60
1	9·24039	71	9·24706	73	10·75294	9·99333	2	59
2	9·24110	71	9·24779	74	10·75221	9·99331	3	58
3	9·24181	72	9·24853	73	10·75147	9·99328	2	57
4	9·24253	71	9·24926	74	10·75074	9·99326	2	56
5	9·24324	71	9·25000	73	10·75000	9·99324	2	55
6	9·24395	71	9·25073	73	10·74927	9·99322	3	54
7	9·24466	70	9·25146	73	10·74854	9·99319	2	53
8	9·24536	71	9·25219	73	10·74781	9·99317	2	52
9	9·24607	70	9·25292	73	10·74708	9·99315	2	51
10	9·24677	71	9·25365	72	10·74635	9·99313	3	50
11	9·24748	70	9·25437	73	10·74563	9·99310	2	49
12	9·24818	70	9·25510	72	10·74490	9·99308	2	48
13	9·24888	70	9·25582	73	10·74418	9·99306	2	47
14	9·24958	70	9·25655	72	10·74345	9·99304	3	46
15	9·25028	70	9·25727	72	10·74273	9·99301	2	45
16	9·25098	70	9·25799	72	10·74201	9·99299	2	44
17	9·25168	69	9·25871	72	10·74129	9·99297	3	43
18	9·25237	70	9·25943	72	10·74057	9·99294	2	42
19	9·25307	69	9·26015	71	10·73985	9·99292	2	41
20	9·25376	69	9·26086	72	10·73914	9·99290	2	40
21	9·25445	69	9·26158	71	10·73842	9·99288	3	39
22	9·25514	69	9·26229	72	10·73771	9·99285	2	38
23	9·25583	69	9·26301	71	10·73699	9·99283	2	37
24	9·25652	69	9·26372	72	10·73628	9·99281	3	36
25	9·25721	69	9·26443	71	10·73557	9·99278	2	35
26	9·25790	68	9·26514	71	10·73486	9·99276	2	34
27	9·25858	69	9·26585	70	10·73415	9·99274	3	33
28	9·25927	68	9·26655	71	10·73345	9·99271	2	32
29	9·25995	68	9·26726	71	10·73274	9·99269	2	31
30	9·26063		9·26797		10·73203	9·99267		30
′	Cosine.		Tangent.		Cotang.	Sine.	D.	′

[10 degrees.] [79 degrees.]

[11 degrees.]   [78 degrees.]

'	D.	Cosine.	Cotang.	Diff.	Tangent.	Diff.	Sine.	'
30	2	9·99119	10·69154	65	9·30846	62	9·29966	30
29	3	9·99117	10·69089	65	9·30911	62	9·30028	31
28	3	9·99114	10·69025	64	9·30975	61	9·30090	32
27	3	9·99112	10·68960	65	9·31040	62	9·30151	33
26	3	9·99109	10·68896	64	9·31104	62	9·30213	34
25	2	9·99106	10·68832	64	9·31168	61	9·30275	35
24	3	9·99104	10·68767	65	9·31233	62	9·30336	36
23	2	9·99101	10·68703	64	9·31297	61	9·30398	37
22	2	9·99099	10·68639	64	9·31361	62	9·30459	38
21	3	9·99096	10·68575	64	9·31425	61	9·30521	39
20	2	9·99093	10·68511	64	9·31489	61	9·30582	40
19	3	9·99091	10·68448	63	9·31552	61	9·30643	41
18	2	9·99088	10·68384	63	9·31616	61	9·30704	42
17	3	9·99086	10·68321	63	9·31679	61	9·30765	43
16	3	9·99083	10·68257	64	9·31743	61	9·30826	44
15	3	9·99080	10·68194	64	9·31806	60	9·30887	45
14	2	9·99078	10·68130	63	9·31870	61	9·30947	46
13	3	9·99075	10·68067	63	9·31933	60	9·31008	47
12	2	9·99072	10·68004	63	9·31996	61	9·31068	48
11	3	9·99070	10·67941	63	9·32059	60	9·31129	49
10	3	9·99067	10·67878	63	9·32122	61	9·31189	50
9	3	9·99064	10·67815	63	9·32185	60	9·31250	51
8	2	9·99062	10·67752	63	9·32248	60	9·31310	52
7	3	9·99059	10·67689	62	9·32311	60	9·31370	53
6	2	9·99056	10·67627	63	9·32373	60	9·31430	54
5	3	9·99054	10·67564	62	9·32436	59	9·31490	55
4	3	9·99051	10·67502	63	9·32498	60	9·31549	56
3	2	9·99048	10·67439	62	9·32561	60	9·31609	57
2	3	9·99046	10·67377	62	9·32623	59	9·31669	58
1	3	9·99043	10·67315	62	9·32685	60	9·31728	59
0		9·99040	10·67253		9·32747		9·31788	60
'		Sine.	Tangent.		Cotang.		Cosine.	'

[11 degrees.]   [78 degrees.]

[11 degrees.]   [78 degrees.]

'	D.	Cosine.	Cotang.	Diff.	Tangent.	Diff.	Sine.	'
0	3	9·99195	10·71135	68	9·28865	65	9·28060	60
1	2	9·99192	10·71067	67	9·28933	65	9·28125	59
2	3	9·99190	10·71000	67	9·29000	64	9·28190	58
3	2	9·99187	10·70933	67	9·29067	65	9·28254	57
4	3	9·99185	10·70866	67	9·29134	65	9·28319	56
5	2	9·99182	10·70799	67	9·29201	64	9·28384	55
6	3	9·99180	10·70732	67	9·29268	64	9·28448	54
7	2	9·99177	10·70665	67	9·29335	65	9·28512	53
8	3	9·99175	10·70598	66	9·29402	64	9·28577	52
9	2	9·99172	10·70532	67	9·29468	64	9·28641	51
10	3	9·99170	10·70465	66	9·29535	64	9·28705	50
11	2	9·99167	10·70399	67	9·29601	64	9·28769	49
12	3	9·99165	10·70332	66	9·29668	63	9·28833	48
13	2	9·99162	10·70266	66	9·29734	64	9·28896	47
14	3	9·99160	10·70200	66	9·29800	64	9·28960	46
15	2	9·99157	10·70134	66	9·29866	63	9·29024	45
16	3	9·99155	10·70068	66	9·29932	63	9·29087	44
17	2	9·99152	10·70002	66	9·29998	64	9·29150	43
18	3	9·99150	10·69936	66	9·30064	63	9·29214	42
19	3	9·99147	10·69870	65	9·30130	63	9·29277	41
20	2	9·99145	10·69805	66	9·30195	63	9·29340	40
21	3	9·99142	10·69739	65	9·30261	63	9·29403	39
22	2	9·99140	10·69674	65	9·30326	63	9·29466	38
23	3	9·99137	10·69609	66	9·30391	62	9·29529	37
24	3	9·99135	10·69543	65	9·30457	63	9·29591	36
25	2	9·99132	10·69478	65	9·30522	63	9·29654	35
26	3	9·99130	10·69413	65	9·30587	62	9·29716	34
27	2	9·99127	10·69348	65	9·30652	63	9·29779	33
28	3	9·99124	10·69283	65	9·30717	62	9·29841	32
29	2	9·99122	10·69218	64	9·30782	63	9·29903	31
30	3	9·99119	10·69154		9·30846		9·29966	30
'		Cosine.	Tangent.		Cotang.		Sine.	'

[11 degrees.]   [78 degrees.]

[12 degrees.]   [77 degrees.]

′	D.	Cosine.	Cotang.	Diff.	Tangent.	Diff.	Sine.	′
30	3	9·98958	10·65424	59	9·34576	57	9·33534	30
29	2	9·98955	10·65365	60	9·34635	57	9·33591	31
28	3	9·98953	10·65305	60	9·34695	57	9·33647	32
27	3	9·98950	10·65245	59	9·34755	57	9·33704	33
26	3	9·98947	10·65186	60	9·34814	57	9·33761	34
25	3	9·98944	10·65126	59	9·34874	56	9·33818	35
24	3	9·98941	10·65067	59	9·34933	57	9·33874	36
23	3	9·98938	10·65008	59	9·34992	56	9·33931	37
22	2	9·98936	10·64949	59	9·35051	56	9·33987	38
21	3	9·98933	10·64889	60	9·35111	56	9·34043	39
20	3	9·98930	10·64830	59	9·35170	57	9·34100	40
19	3	9·98927	10·64771	59	9·35229	56	9·34156	41
18	3	9·98924	10·64712	59	9·35288	56	9·34212	42
17	3	9·98921	10·64653	59	9·35347	56	9·34268	43
16	2	9·98919	10·64595	58	9·35405	56	9·34324	44
15	3	9·98916	10·64536	59	9·35464	56	9·34380	45
14	3	9·98913	10·64477	58	9·35523	56	9·34436	46
13	3	9·98910	10·64419	58	9·35581	55	9·34491	47
12	3	9·98907	10·64360	59	9·35640	56	9·34547	48
11	3	9·98904	10·64302	58	9·35698	55	9·34602	49
10	3	9·98901	10·64243	59	9·35757	56	9·34658	50
9	3	9·98898	10·64185	58	9·35815	55	9·34713	51
8	3	9·98896	10·64127	58	9·35873	56	9·34769	52
7	3	9·98893	10·64069	58	9·35931	55	9·34824	53
6	3	9·98890	10·64011	58	9·35989	55	9·34879	54
5	3	9·98887	10·63953	58	9·36047	55	9·34934	55
4	3	9·98884	10·63895	57	9·36105	55	9·34989	56
3	3	9·98881	10·63837	58	9·36163	55	9·35044	57
2	3	9·98878	10·63779	58	9·36221	55	9·35099	58
1	3	9·98875	10·63721	57	9·36279	55	9·35154	59
0	3	9·98872	10·63664		9·36336		9·35209	60
′		Sine.	Tangent.		Cotang.		Cosine.	′

113

[12 degrees.]   [77 degrees.]

′	D.	Cosine.	Cotang.	Diff.	Tangent.	Diff.	Sine.	′
0	2	9·99040	10·67253	63	9·32747	59	9·31788	60
1	3	9·99038	10·67190	62	9·32810	59	9·31847	59
2	3	9·99035	10·67128	61	9·32872	60	9·31907	58
3	2	9·99032	10·67067	62	9·32933	59	9·31966	57
4	3	9·99030	10·67005	62	9·32995	59	9·32025	56
5	3	9·99027	10·66943	62	9·33057	59	9·32084	55
6	3	9·99024	10·66881	61	9·33119	59	9·32143	54
7	2	9·99022	10·66820	62	9·33180	59	9·32202	53
8	3	9·99019	10·66758	61	9·33242	59	9·32261	52
9	3	9·99016	10·66697	62	9·33303	58	9·32319	51
10	3	9·99013	10·66635	61	9·33365	59	9·32378	50
11	2	9·99011	10·66574	61	9·33426	59	9·32437	49
12	3	9·99008	10·66513	61	9·33487	58	9·32495	48
13	3	9·99005	10·66452	61	9·33548	58	9·32553	47
14	3	9·99002	10·66391	61	9·33609	59	9·32612	46
15	3	9·99000	10·66330	61	9·33670	58	9·32670	45
16	3	9·98997	10·66269	61	9·33731	58	9·32728	44
17	2	9·98994	10·66208	61	9·33792	58	9·32786	43
18	3	9·98991	10·66147	60	9·33853	58	9·32844	42
19	3	9·98989	10·66087	61	9·33913	58	9·32902	41
20	3	9·98986	10·66026	60	9·33974	58	9·32960	40
21	3	9·98983	10·65966	61	9·34034	58	9·33018	39
22	2	9·98980	10·65905	60	9·34095	57	9·33075	38
23	3	9·98978	10·65845	60	9·34155	58	9·33133	37
24	3	9·98975	10·65785	61	9·34215	57	9·33190	36
25	3	9·98972	10·65724	60	9·34276	58	9·33248	35
26	3	9·98969	10·65664	60	9·34336	57	9·33305	34
27	2	9·98967	10·65604	60	9·34396	57	9·33362	33
28	3	9·98964	10·65544	60	9·34456	58	9·33420	32
29	3	9·98961	10·65484	60	9·34516	57	9·33477	31
30	3	9·98958	10·65424		9·34576		9·33534	30
′		Sine.	Tangent.		Cotang.		Cosine.	′

[13 degrees.]   [76 degrees.]

'	Sine.	Diff.	Tangent.	Diff.	Cotang.	Cosine.	D.	'
30	9·36819	52	9·38035	56	10·61965	9·98783	3	30
31	9·36871	52	9·38091	56	10·61909	9·98780	3	29
32	9·36924	53	9·38147	55	10·61853	9·98777	3	28
33	9·36976	52	9·38202	55	10·61798	9·98774	3	27
34	9·37028	53	9·38257	56	10·61743	9·98771	3	26
35	9·37081	52	9·38313	55	10·61687	9·98768	3	25
36	9·37133	52	9·38368	55	10·61632	9·98765	3	24
37	9·37185	52	9·38423	56	10·61577	9·98762	3	23
38	9·37237	52	9·38479	55	10·61521	9·98759	3	22
39	9·37289	52	9·38534	55	10·61466	9·98756	4	21
40	9·37341	52	9·38589	55	10·61411	9·98753	3	20
41	9·37393	52	9·38644	55	10·61356	9·98750	3	19
42	9·37445	52	9·38699	55	10·61301	9·98746	3	18
43	9·37497	52	9·38754	54	10·61246	9·98743	3	17
44	9·37549	51	9·38808	55	10·61192	9·98740	3	16
45	9·37600	52	9·38863	55	10·61137	9·98737	3	15
46	9·37652	51	9·38918	54	10·61082	9·98734	3	14
47	9·37703	52	9·38972	55	10·61028	9·98731	3	13
48	9·37755	51	9·39027	55	10·60973	9·98728	3	12
49	9·37806	52	9·39082	54	10·60918	9·98725	3	11
50	9·37858	51	9·39136	54	10·60864	9·98722	3	10
51	9·37909	51	9·39190	55	10·60810	9·98719	4	9
52	9·37960	51	9·39245	54	10·60755	9·98715	3	8
53	9·38011	51	9·39299	54	10·60701	9·98712	3	7
54	9·38062	51	9·39353	54	10·60647	9·98709	3	6
55	9·38113	51	9·39407	54	10·60593	9·98706	3	5
56	9·38164	51	9·39461	54	10·60539	9·98703	3	4
57	9·38215	51	9·39515	54	10·60485	9·98700	3	3
58	9·38266	51	9·39569	54	10·60431	9·98697	3	2
59	9·38317	51	9·39623	54	10·60377	9·98694	4	1
60	9·38368		9·39677		10·60323	9·98690		0
'	Cosine.		Cotang.		Tangent.	Sine.		'

[76 degrees.]

114

[13 degrees.]   [76 degrees.]

'	Sine.	Diff.	Tangent.	Diff.	Cotang.	Cosine.	D.	'
0	9·35209	54	9·36336	58	10·63664	9·98872	3	60
1	9·35263	55	9·36394	58	10·63606	9·98869	2	59
2	9·35318	55	9·36452	57	10·63548	9·98867	3	58
3	9·35373	54	9·36509	57	10·63491	9·98864	3	57
4	9·35427	54	9·36566	58	10·63434	9·98861	3	56
5	9·35481	55	9·36624	57	10·63376	9·98858	3	55
6	9·35536	54	9·36681	57	10·63319	9·98855	3	54
7	9·35590	54	9·36738	57	10·63262	9·98852	3	53
8	9·35644	54	9·36795	57	10·63205	9·98849	3	52
9	9·35698	54	9·36852	57	10·63148	9·98846	3	51
10	9·35752	54	9·36909	57	10·63091	9·98843	3	50
11	9·35806	54	9·36966	57	10·63034	9·98840	3	49
12	9·35860	54	9·37023	57	10·62977	9·98837	3	48
13	9·35914	54	9·37080	57	10·62920	9·98834	3	47
14	9·35968	54	9·37137	56	10·62863	9·98831	3	46
15	9·36022	53	9·37193	57	10·62807	9·98828	3	45
16	9·36075	54	9·37250	56	10·62750	9·98825	3	44
17	9·36129	53	9·37306	57	10·62694	9·98822	3	43
18	9·36182	54	9·37363	56	10·62637	9·98819	3	42
19	9·36236	53	9·37419	57	10·62581	9·98816	3	41
20	9·36289	53	9·37476	56	10·62524	9·98813	3	40
21	9·36342	53	9·37532	56	10·62468	9·98810	3	39
22	9·36395	54	9·37588	56	10·62412	9·98807	3	38
23	9·36449	53	9·37644	56	10·62356	9·98804	3	37
24	9·36502	53	9·37700	56	10·62300	9·98801	3	36
25	9·36555	53	9·37756	56	10·62244	9·98798	3	35
26	9·36608	52	9·37812	56	10·62188	9·98795	3	34
27	9·36660	53	9·37868	56	10·62132	9·98792	3	33
28	9·36713	53	9·37924	56	10·62076	9·98789	3	32
29	9·36766	53	9·37980	55	10·62020	9·98786	3	31
30	9·36819		9·38035		10·61965	9·98783	3	30
'	Cosine.		Cotang.		Tangent.	Sine.		'

[76 degrees.]

[14 degrees.]   **[75 degrees.]**

'	D.	Cosine.	Cotang.	Diff.	Tangent.	Diff.	Sine.	'
30	3	9·98594	10·58734	52	9·41266	49	9·39860	30
29	3	9·98591	10·58682	52	9·41318	49	9·39909	31
28	4	9·98588	10·58630	52	9·41370	48	9·39958	32
27	3	9·98584	10·58578	52	9·41422	49	9·40006	33
26	3	9·98581	10·58526	52	9·41474	49	9·40055	34
25	4	9·98578	10·58474	52	9·41526	48	9·40103	35
24	3	9·98574	10·58422	51	9·41578	49	9·40152	36
23	3	9·98571	10·58371	52	9·41629	48	9·40200	37
22	3	9·98568	10·58319	52	9·41681	49	9·40249	38
21	4	9·98565	10·58267	51	9·41733	48	9·40297	39
20	3	9·98561	10·58216	52	9·41784	49	9·40346	40
19	3	9·98558	10·58164	51	9·41836	48	9·40394	41
18	4	9·98555	10·58113	52	9·41887	48	9·40442	42
17	3	9·98551	10·58061	51	9·41939	48	9·40490	43
16	3	9·98548	10·58010	51	9·41990	48	9·40538	44
15	3	9·98545	10·57959	52	9·42041	48	9·40586	45
14	4	9·98541	10·57907	51	9·42093	48	9·40634	46
13	3	9·98538	10·57856	51	9·42144	48	9·40682	47
12	3	9·98535	10·57805	51	9·42195	48	9·40730	48
11	4	9·98531	10·57754	51	9·42246	47	9·40778	49
10	3	9·98528	10·57703	51	9·42297	48	9·40825	50
9	3	9·98525	10·57652	51	9·42348	48	9·40873	51
8	4	9·98521	10·57601	51	9·42399	47	9·40921	52
7	3	9·98518	10·57550	51	9·42450	48	9·40968	53
6	3	9·98515	10·57499	51	9·42501	47	9·41016	54
5	4	9·98511	10·57448	51	9·42552	48	9·41063	55
4	3	9·98508	10·57397	50	9·42603	47	9·41111	56
3	3	9·98505	10·57347	51	9·42653	47	9·41158	57
2	4	9·98501	10·57296	51	9·42704	47	9·41205	58
1	3	9·98498	10·57245	50	9·42755	48	9·41252	59
0	4	9·98494	10·57195	50	9·42805		9·41300	60
		Sine.	Tangent.		Cotang.		Cosine.	

115

[14 degrees.]   **[75 degrees.]**

'	Sine.	Diff.	Tangent.	Diff.	Cotang.	D.	Cosine.	'
0	9·38368	50	9·39677	54	10·60323	3	9·98690	60
1	9·38418	51	9·39731	54	10·60269	3	9·98687	59
2	9·38469	50	9·39785	53	10·60215	3	9·98684	58
3	9·38519	51	9·39838	54	10·60162	3	9·98681	57
4	9·38570	50	9·39892	53	10·60108	3	9·98678	56
5	9·38620	50	9·39945	54	10·60055	3	9·98675	55
6	9·38670	51	9·39999	53	10·60001	3	9·98671	54
7	9·38721	50	9·40052	54	10·59948	3	9·98668	53
8	9·38771	50	9·40106	53	10·59894	3	9·98665	52
9	9·38821	50	9·40159	53	10·59841	3	9·98662	51
10	9·38871	50	9·40212	54	10·59788	3	9·98659	50
11	9·38921	50	9·40266	53	10·59734	3	9·98656	49
12	9·38971	50	9·40319	53	10·59681	4	9·98652	48
13	9·39021	50	9·40372	53	10·59628	3	9·98649	47
14	9·39071	50	9·40425	53	10·59575	3	9·98646	46
15	9·39121	49	9·40478	53	10·59522	3	9·98643	45
16	9·39170	50	9·40531	53	10·59469	4	9·98640	44
17	9·39220	50	9·40584	52	10·59416	3	9·98636	43
18	9·39270	49	9·40636	53	10·59364	3	9·98633	42
19	9·39319	50	9·40689	53	10·59311	3	9·98630	41
20	9·39369	49	9·40742	53	10·59258	4	9·98627	40
21	9·39418	49	9·40795	52	10·59205	3	9·98623	39
22	9·39467	50	9·40847	53	10·59153	3	9·98620	38
23	9·39517	49	9·40900	52	10·59100	3	9·98617	37
24	9·39566	49	9·40952	53	10·59048	4	9·98614	36
25	9·39615	49	9·41005	52	10·58995	3	9·98610	35
26	9·39664	49	9·41057	52	10·58943	3	9·98607	34
27	9·39713	49	9·41109	52	10·58891	3	9·98604	33
28	9·39762	49	9·41161	53	10·58839	3	9·98601	32
29	9·39811	49	9·41214	52	10·58786	3	9·98597	31
30	9·39860		9·41266		10·58734	3	9·98594	30
	Cosine.		Cotang.		Tangent.		Sine.	

[15 degrees.]   [74 degrees.]

´	D.	Cosine.	Cotang.	Diff.	Tangent.	Diff.	Sine.	´
30	3	9·98391	10·55701	49	9·44299	45	9·42690	30
29	4	9·98388	10·55652	49	9·44348	45	9·42735	31
28	3	9·98384	10·55603	49	9·44397	46	9·42781	32
27	4	9·98381	10·55554	49	9·44446	45	9·42826	33
26	4	9·98377	10·55505	49	9·44495	46	9·42872	34
25	3	9·98373	10·55456	48	9·44544	45	9·42917	35
24	4	9·98370	10·55408	49	9·44592	45	9·42962	36
23	3	9·98366	10·55359	49	9·44641	46	9·43008	37
22	4	9·98363	10·55310	48	9·44690	45	9·43053	38
21	3	9·98359	10·55262	49	9·44738	45	9·43098	39
20	3	9·98356	10·55213	49	9·44787	45	9·43143	40
19	3	9·98352	10·55164	48	9·44836	45	9·43188	41
18	4	9·98349	10·55116	48	9·44884	45	9·43233	42
17	3	9·98345	10·55067	49	9·44933	45	9·43278	43
16	4	9·98342	10·55019	48	9·44981	44	9·43323	44
15	4	9·98338	10·54971	49	9·45029	45	9·43367	45
14	3	9·98334	10·54922	48	9·45078	45	9·43412	46
13	4	9·98331	10·54874	48	9·45126	45	9·43457	47
12	3	9·98327	10·54826	48	9·45174	45	9·43502	48
11	3	9·98324	10·54778	49	9·45222	44	9·43546	49
10	3	9·98320	10·54729	48	9·45271	45	9·43591	50
9	4	9·98317	10·54681	48	9·45319	44	9·43635	51
8	3	9·98313	10·54633	48	9·45367	45	9·43680	52
7	4	9·98309	10·54585	48	9·45415	44	9·43724	53
6	4	9·98306	10·54537	48	9·45463	45	9·43769	54
5	3	9·98302	10·54489	48	9·45511	44	9·43813	55
4	4	9·98299	10·54441	47	9·45559	44	9·43857	56
3	3	9·98295	10·54394	48	9·45606	44	9·43901	57
2	4	9·98291	10·54346	48	9·45654	45	9·43946	58
1	3	9·98288	10·54298	48	9·45702	44	9·43990	59
0	4	9·98284	10·54250	48	9·45750	44	9·44034	60
´		Sine.	Tangent.	Diff.	Cotang.	Diff.	Cosine.	´

[15 degrees.]   [74 degrees.]

´	D.	Cosine.	Cotang.	Diff.	Tangent.	Diff.	Sine.	´
60	3	9·98494	10·57195	51	9·42805	47	9·41300	0
59	3	9·98491	10·57144	50	9·42856	47	9·41347	1
58	4	9·98488	10·57094	51	9·42906	47	9·41394	2
57	3	9·98484	10·57043	50	9·42957	47	9·41441	3
56	4	9·98481	10·56993	50	9·43007	47	9·41488	4
55	3	9·98477	10·56943	51	9·43057	47	9·41535	5
54	3	9·98474	10·56892	50	9·43108	47	9·41582	6
53	3	9·98471	10·56842	50	9·43158	46	9·41628	7
52	4	9·98467	10·56792	50	9·43208	47	9·41675	8
51	3	9·98464	10·56742	50	9·43258	47	9·41722	9
50	4	9·98460	10·56692	50	9·43308	46	9·41762	10
49	3	9·98457	10·56642	50	9·43358	47	9·41815	11
48	4	9·98453	10·56592	50	9·43408	46	9·41861	12
47	3	9·98450	10·56542	50	9·43458	47	9·41908	13
46	4	9·98447	10·56492	50	9·43508	46	9·41954	14
45	3	9·98443	10·56442	49	9·43558	47	9·42001	15
44	4	9·98440	10·56393	50	9·43607	46	9·42047	16
43	3	9·98436	10·56343	50	9·43657	46	9·42093	17
42	4	9·98433	10·56293	49	9·43707	47	9·42140	18
41	4	9·98429	10·56244	50	9·43756	46	9·42186	19
40	3	9·98426	10·56194	49	9·43806	46	9·42232	20
39	4	9·98422	10·56145	50	9·43855	46	9·42278	21
38	3	9·98419	10·56095	49	9·43905	46	9·42324	22
37	4	9·98415	10·56046	50	9·43954	46	9·42370	23
36	3	9·98412	10·55996	49	9·44004	45	9·42416	24
35	3	9·98409	10·55947	49	9·44053	46	9·42461	25
34	4	9·98405	10·55898	49	9·44102	46	9·42507	26
33	3	9·98402	10·55849	50	9·44151	46	9·42553	27
32	4	9·98398	10·55799	49	9·44201	45	9·42599	28
31	3	9·98395	10·55750	49	9·44250	45	9·42644	29
30	4	9·98391	10·55701		9·44299	46	9·42690	30
´		Sine.	Tangent.	Diff.	Cotang.	Diff.	Cosine.	´

[16 degrees.]

'	Sine.	Diff.	Tangent.	Diff.	Cotang.	Cosine.	D.	'
30	9·45334	43	9·47160	47	10·52840	9·98174	4	30
31	9·45377	42	9·47207	46	10·52793	9·98170	4	29
32	9·45419	43	9·47253	46	10·52747	9·98166	4	28
33	9·45462	42	9·47299	47	10·52701	9·98162	3	27
34	9·45504	43	9·47346	46	10·52654	9·98159	4	26
35	9·45547	42	9·47392	46	10·52608	9·98155	4	25
36	9·45589	43	9·47438	46	10·52562	9·98151	4	24
37	9·45632	42	9·47484	46	10·52516	9·98147	3	23
38	9·45674	42	9·47530	46	10·52470	9·98144	4	22
39	9·45716	42	9·47576	46	10·52424	9·98140	4	21
40	9·45758	43	9·47622	46	10·52378	9·98136	4	20
41	9·45801	42	9·47668	46	10·52332	9·98132	3	19
42	9·45843	42	9·47714	46	10·52286	9·98129	4	18
43	9·45885	42	9·47760	46	10·52240	9·98125	4	17
44	9·45927	42	9·47806	46	10·52194	9·98121	4	16
45	9·45969	42	9·47852	45	10·52148	9·98117	3	15
46	9·46011	42	9·47897	46	10·52103	9·98113	4	14
47	9·46053	42	9·47943	46	10·52057	9·98110	3	13
48	9·46095	41	9·47989	46	10·52011	9·98106	4	12
49	9·46136	42	9·48035	45	10·51965	9·98102	4	11
50	9·46178	42	9·48080	46	10·51920	9·98098	4	10
51	9·46220	42	9·48126	45	10·51874	9·98094	4	9
52	9·46262	41	9·48171	46	10·51829	9·98090	3	8
53	9·46303	42	9·48217	45	10·51783	9·98087	4	7
54	9·46345	41	9·48262	45	10·51738	9·98083	4	6
55	9·46386	42	9·48307	46	10·51693	9·98079	4	5
56	9·46428	41	9·48353	45	10·51647	9·98075	4	4
57	9·46469	42	9·48398	45	10·51602	9·98071	4	3
58	9·46511	41	9·48443	46	10·51557	9·98067	4	2
59	9·46552	42	9·48489	45	10·51511	9·98063	4	1
60	9·46594		9·48534		10·51466	9·98060	3	0
'	Cosine.		Tangent.		Cotang.	Sine.		'

[73 degrees.]

117

[16 degrees.]

'	Sine.	Diff.	Tangent.	Diff.	Cotang.	Cosine.	D.	'
0	9·44034	44	9·45750	47	10·54250	9·98284	3	60
1	9·44078	44	9·45797	47	10·54203	9·98281	4	59
2	9·44122	44	9·45845	48	10·54155	9·98277	4	58
3	9·44166	44	9·45892	47	10·54108	9·98273	3	57
4	9·44210	43	9·45940	48	10·54060	9·98270	4	56
5	9·44253	44	9·45987	47	10·54013	9·98266	4	55
6	9·44297	44	9·46035	48	10·53965	9·98262	3	54
7	9·44341	44	9·46082	47	10·53918	9·98259	4	53
8	9·44385	43	9·46130	48	10·53870	9·98255	4	52
9	9·44428	44	9·46177	47	10·53823	9·98251	3	51
10	9·44472	44	9·46224	47	10·53776	9·98248	4	50
11	9·44516	43	9·46271	48	10·53729	9·98244	4	49
12	9·44559	43	9·46319	47	10·53681	9·98240	3	48
13	9·44602	44	9·46366	47	10·53634	9·98237	4	47
14	9·44646	43	9·46413	47	10·53587	9·98233	4	46
15	9·44689	44	9·46460	47	10·53540	9·98229	3	45
16	9·44733	43	9·46507	47	10·53493	9·98226	4	44
17	9·44776	43	9·46554	47	10·53446	9·98222	4	43
18	9·44819	43	9·46601	47	10·53399	9·98218	4	42
19	9·44862	43	9·46648	47	10·53352	9·98215	3	41
20	9·44905	43	9·46694	46	10·53306	9·98211	4	40
21	9·44948	44	9·46741	47	10·53259	9·98207	3	39
22	9·44992	43	9·46788	47	10·53212	9·98204	4	38
23	9·45035	42	9·46835	46	10·53165	9·98200	4	37
24	9·45077	43	9·46881	47	10·53119	9·98196	4	36
25	9·45120	43	9·46928	47	10·53072	9·98192	3	35
26	9·45163	43	9·46975	46	10·53025	9·98189	4	34
27	9·45206	43	9·47021	47	10·52979	9·98185	4	33
28	9·45249	43	9·47068	46	10·52932	9·98181	4	32
29	9·45292	42	9·47114	46	10·52886	9·98177	3	31
30	9·45334		9·47160		10·52840	9·98174		30
'	Cosine.		Tangent.		Cotang.	Sine.		'

[73 degrees.]

[17 degrees.] [72 degrees.]

'	D.	Cosine.	Cotang.	Diff.	Tangent.	Diff.	Sine.	'
30	4	9·97942	10·50128	44	9·49872	40	9·47814	30
31	4	9·97938	10·50084	44	9·49916	40	9·47854	29
32	4	9·97934	10·50040	44	9·49960	40	9·47894	28
33	4	9·97930	10·49996	44	9·50004	40	9·47934	27
34	4	9·97926	10·49952	44	9·50048	40	9·47974	26
35	4	9·97922	10·49908	44	9·50092	40	9·48014	25
36	4	9·97918	10·49864	44	9·50136	40	9·48054	24
37	4	9·97914	10·49820	43	9·50180	40	9·48094	23
38	4	9·97910	10·49777	44	9·50223	39	9·48133	22
39	4	9·97906	10·49733	44	9·50267	40	9·48173	21
40	4	9·97902	10·49689	44	9·50311	40	9·48213	20
41	4	9·97898	10·49645	43	9·50355	39	9·48252	19
42	4	9·97894	10·49602	44	9·50398	40	9·48292	18
43	4	9·97890	10·49558	43	9·50442	40	9·48332	17
44	4	9·97886	10·49515	44	9·50485	39	9·48371	16
45	4	9·97882	10·49471	43	9·50529	40	9·48411	15
46	4	9·97878	10·49428	44	9·50572	39	9·48450	14
47	5	9·97874	10·49384	43	9·50616	39	9·48490	13
48	4	9·97870	10·49341	44	9·50659	39	9·48529	12
49	4	9·97866	10·49297	43	9·50703	39	9·48568	11
50	4	9·97861	10·49254	43	9·50746	40	9·48607	10
51	4	9·97857	10·49211	44	9·50789	39	9·48647	9
52	4	9·97853	10·49167	43	9·50833	39	9·48686	8
53	4	9·97849	10·49124	43	9·50876	39	9·48725	7
54	4	9·97845	10·49081	44	9·50919	40	9·48764	6
55	4	9·97841	10·49038	43	9·50962	39	9·48803	5
56	4	9·97837	10·48995	43	9·51005	39	9·48842	4
57	4	9·97833	10·48952	44	9·51048	39	9·48881	3
58	4	9·97829	10·48908	43	9·51092	39	9·48920	2
59	4	9·97825	10·48865	43	9·51135	39	9·48959	1
60	4	9·97821	10·48822		9·51178		9·48998	0
'		Sine.	Tangent.		Cotang.		Cosine.	'

[72 degrees.]

118

[17 degrees.] [72 degrees.]

'	D.	Cosine.	Cotang.	Diff.	Tangent.	Diff.	Sine.	'
0	4	9·98060	10·51466	45	9·48534	41	9·46594	60
1	4	9·98056	10·51421	45	9·48579	41	9·46635	59
2	4	9·98052	10·51376	45	9·48624	41	9·46676	58
3	4	9·98048	10·51331	45	9·48669	41	9·46717	57
4	4	9·98044	10·51286	45	9·48714	42	9·46758	56
5	4	9·98040	10·51241	45	9·48759	41	9·46800	55
6	4	9·98036	10·51196	45	9·48804	41	9·46841	54
7	4	9·98032	10·51151	45	9·48849	41	9·46882	53
8	4	9·98029	10·51106	45	9·48894	41	9·46923	52
9	4	9·98025	10·51061	45	9·48939	41	9·46964	51
10	4	9·98021	10·51016	45	9·48984	41	9·47005	50
11	4	9·98017	10·50971	44	9·49029	41	9·47045	49
12	4	9·98013	10·50927	45	9·49073	41	9·47086	48
13	4	9·98009	10·50882	45	9·49118	41	9·47127	47
14	4	9·98005	10·50837	44	9·49163	41	9·47168	46
15	4	9·98001	10·50793	45	9·49207	41	9·47209	45
16	4	9·97997	10·50748	44	9·49252	40	9·47249	44
17	4	9·97993	10·50704	44	9·49296	41	9·47290	43
18	4	9·97989	10·50659	45	9·49341	40	9·47330	42
19	3	9·97986	10·50615	44	9·49385	41	9·47371	41
20	4	9·97982	10·50570	44	9·49430	40	9·47411	40
21	4	9·97978	10·50526	45	9·49474	41	9·47452	39
22	4	9·97974	10·50481	44	9·49519	40	9·47492	38
23	4	9·97970	10·50437	44	9·49563	41	9·47533	37
24	4	9·97966	10·50393	45	9·49607	40	9·47573	36
25	4	9·97962	10·50348	44	9·49652	41	9·47613	35
26	4	9·97958	10·50304	44	9·49696	40	9·47654	34
27	4	9·97954	10·50260	44	9·49740	40	9·47694	33
28	4	9·97950	10·50216	44	9·49784	40	9·47734	32
29	4	9·97946	10·50172	44	9·49828	40	9·47774	31
30	4	9·97942	10·50128		9·49872		9·47814	30
'		Cosine.	Tangent.		Cotang.		Sine.	'

[18 degrees.]   [71 degrees.]

′	Sine	Diff.	Tangent	Diff.	Cotang.	Cosine	D.	′
30	9·50148	37	9·52452	42	10·47548	9·97696	5	30
31	9·50185	38	9·52494	42	10·47506	9·97691	5	29
32	9·50223	38	9·52536	42	10·47464	9·97687	4	28
33	9·50261	37	9·52578	42	10·47422	9·97683	4	27
34	9·50298	38	9·52620	41	10·47380	9·97679	4	26
35	9·50336	38	9·52661	42	10·47339	9·97674	4	25
36	9·50374	37	9·52703	42	10·47297	9·97670	4	24
37	9·50411	38	9·52745	42	10·47255	9·97666	4	23
38	9·50449	37	9·52787	42	10·47213	9·97662	5	22
39	9·50486	37	9·52829	41	10·47171	9·97657	4	21
40	9·50523	38	9·52870	42	10·47130	9·97653	4	20
41	9·50561	37	9·52912	41	10·47088	9·97649	4	19
42	9·50598	37	9·52953	42	10·47047	9·97645	5	18
43	9·50635	38	9·52995	42	10·47005	9·97640	4	17
44	9·50673	37	9·53037	41	10·46963	9·97636	4	16
45	9·50710	37	9·53078	42	10·46922	9·97632	4	15
46	9·50747	37	9·53120	41	10·46880	9·97628	5	14
47	9·50784	37	9·53161	41	10·46839	9·97623	4	13
48	9·50821	37	9·53202	42	10·46798	9·97619	4	12
49	9·50858	38	9·53244	41	10·46756	9·97615	5	11
50	9·50896	37	9·53285	42	10·46715	9·97610	4	10
51	9·50933	37	9·53327	41	10·46673	9·97606	4	9
52	9·50970	37	9·53368	41	10·46632	9·97602	5	8
53	9·51007	36	9·53409	41	10·46591	9·97597	4	7
54	9·51043	37	9·53450	42	10·46550	9·97593	4	6
55	9·51080	37	9·53492	41	10·46508	9·97589	5	5
56	9·51117	37	9·53533	41	10·46467	9·97584	4	4
57	9·51154	37	9·53574	41	10·46426	9·97580	4	3
58	9·51191	36	9·53615	41	10·46385	9·97576	5	2
59	9·51227	37	9·53656	41	10·46344	9·97571	4	1
60	9·51264		9·53697		10·46303	9·97567	4	0
	Cosine.		Cotang.		Tangent.	Sine.		′

[71 degrees.]

119

[18 degrees.]   [71 degrees.]

′	Sine	Diff.	Tangent	Diff.	Cotang.	Cosine	D.	′
0	9·48998	39	9·51178	43	10·48822	9·97821	4	60
1	9·49037	39	9·51221	43	10·48779	9·97817	5	59
2	9·49076	39	9·51264	42	10·48736	9·97812	4	58
3	9·49115	38	9·51306	43	10·48694	9·97808	4	57
4	9·49153	39	9·51349	43	10·48651	9·97804	4	56
5	9·49192	39	9·51392	43	10·48608	9·97800	4	55
6	9·49231	38	9·51435	43	10·48565	9·97796	4	54
7	9·49269	39	9·51478	42	10·48522	9·97792	4	53
8	9·49308	39	9·51520	43	10·48480	9·97788	4	52
9	9·49347	38	9·51563	43	10·48437	9·97784	5	51
10	9·49385	39	9·51606	42	10·48394	9·97779	4	50
11	9·49424	38	9·51648	43	10·48352	9·97775	4	49
12	9·49462	38	9·51691	43	10·48309	9·97771	4	48
13	9·49500	39	9·51734	42	10·48266	9·97767	4	47
14	9·49539	38	9·51776	43	10·48224	9·97763	4	46
15	9·49577	38	9·51819	42	10·48181	9·97759	5	45
16	9·49615	39	9·51861	42	10·48139	9·97754	4	44
17	9·49654	38	9·51903	43	10·48097	9·97750	4	43
18	9·49692	38	9·51946	42	10·48054	9·97746	4	42
19	9·49730	38	9·51988	43	10·48012	9·97742	4	41
20	9·49768	38	9·52031	42	10·47969	9·97738	4	40
21	9·49806	38	9·52073	42	10·47927	9·97734	5	39
22	9·49844	38	9·52115	42	10·47885	9·97729	4	38
23	9·49882	38	9·52157	43	10·47843	9·97725	4	37
24	9·49920	38	9·52200	42	10·47800	9·97721	4	36
25	9·49958	38	9·52242	42	10·47758	9·97717	4	35
26	9·49996	38	9·52284	42	10·47716	9·97713	5	34
27	9·50034	38	9·52326	42	10·47674	9·97708	4	33
28	9·50072	38	9·52368	42	10·47632	9·97704	4	32
29	9·50110	38	9·52410	42	10·47590	9·97700	4	31
30	9·50148		9·52452		10·47548	9·97696		30
	Cosine.		Cotang.		Tangent.	Sine.		′

[71 degrees.]

[19 degrees.]   **[70 degrees.]**

′	D.	Cosine.	Cotang.	Diff.	Tangent.	Diff.	Sine.	′
30	5	9·97435	10·45085	40	9·54915	35	9·52350	30
31	5	9·97430	10·45045	40	9·54955	36	9·52385	29
32	5	9·97426	10·45005	40	9·54995	35	9·52421	28
33	5	9·97421	10·44965	40	9·55035	36	9·52456	27
34	5	9·97417	10·44925	40	9·55075	35	9·52492	26
35	5	9·97412	10·44885	40	9·55115	36	9·52527	25
36	5	9·97408	10·44845	40	9·55155	35	9·52563	24
37	5	9·97403	10·44805	40	9·55195	36	9·52598	23
38	4	9·97399	10·44765	40	9·55235	35	9·52634	22
39	5	9·97394	10·44725	40	9·55275	36	9·52669	21
40	4	9·97390	10·44685	40	9·55315	35	9·52705	20
41	5	9·97385	10·44645	40	9·55355	35	9·52740	19
42	5	9·97381	10·44605	40	9·55395	35	9·52775	18
43	5	9·97376	10·44566	39	9·55434	36	9·52811	17
44	4	9·97372	10·44526	40	9·55474	35	9·52846	16
45	5	9·97367	10·44486	40	9·55514	35	9·52881	15
46	5	9·97363	10·44446	40	9·55554	35	9·52916	14
47	5	9·97358	10·44407	39	9·55593	35	9·52951	13
48	4	9·97353	10·44367	40	9·55633	35	9·52986	12
49	5	9·97349	10·44327	40	9·55673	35	9·53021	11
50	4	9·97344	10·44288	39	9·55712	36	9·53056	10
51	5	9·97340	10·44248	40	9·55752	34	9·53092	9
52	5	9·97335	10·44209	39	9·55791	35	9·53126	8
53	4	9·97331	10·44169	40	9·55831	35	9·53161	7
54	5	9·97326	10·44130	39	9·55870	35	9·53196	6
55	5	9·97322	10·44090	40	9·55910	35	9·53231	5
56	5	9·97317	10·44051	39	9·55949	35	9·53266	4
57	5	9·97312	10·44011	39	9·55989	36	9·53301	3
58	4	9·97308	10·43972	39	9·56028	34	9·53336	2
59	4	9·97303	10·43933	39	9·56067	34	9·53370	1
60	4	9·97299	10·43893		9·56107	35	9·53405	0
′		Sine.	Tangent.		Cotang.		Cosine.	′

[19 degrees.]   **[70 degrees.]**

[19 degrees.]   **[70 degrees.]**

′	D.	Cosine.	Cotang.	Diff.	Tangent.	Diff.	Sine.	′
0	4	9·97567	10·46303	41	9·53697	37	9·51264	60
1	5	9·97563	10·46262	41	9·53738	37	9·51301	59
2	5	9·97558	10·46221	41	9·53779	37	9·51338	58
3	4	9·97554	10·46180	41	9·53820	36	9·51374	57
4	5	9·97550	10·46139	41	9·53861	37	9·51411	56
5	4	9·97545	10·46098	41	9·53902	36	9·51447	55
6	5	9·97541	10·46057	41	9·53943	37	9·51484	54
7	4	9·97536	10·46016	41	9·53984	36	9·51520	53
8	5	9·97532	10·45975	40	9·54025	37	9·51557	52
9	5	9·97528	10·45935	41	9·54065	36	9·51593	51
10	4	9·97523	10·45894	41	9·54106	37	9·51629	50
11	5	9·97519	10·45853	40	9·54147	36	9·51666	49
12	4	9·97515	10·45813	41	9·54187	36	9·51702	48
13	5	9·97510	10·45772	41	9·54228	36	9·51738	47
14	4	9·97506	10·45731	40	9·54269	37	9·51774	46
15	5	9·97501	10·45691	41	9·54309	36	9·51811	45
16	5	9·97497	10·45650	40	9·54350	36	9·51847	44
17	4	9·97492	10·45610	41	9·54390	36	9·51883	43
18	5	9·97488	10·45569	40	9·54431	36	9·51919	42
19	5	9·97484	10·45529	41	9·54471	36	9·51955	41
20	4	9·97479	10·45488	40	9·54512	36	9·51991	40
21	5	9·97475	10·45448	41	9·54552	36	9·52027	39
22	5	9·97470	10·45407	40	9·54593	36	9·52063	38
23	4	9·97466	10·45367	40	9·54633	36	9·52099	37
24	5	9·97461	10·45327	41	9·54673	36	9·52135	36
25	5	9·97457	10·45286	40	9·54714	36	9·52171	35
26	4	9·97453	10·45246	40	9·54754	35	9·52207	34
27	4	9·97448	10·45206	41	9·54794	36	9·52242	33
28	5	9·97444	10·45165	40	9·54835	36	9·52278	32
29	4	9·97439	10·45125	40	9·54875	36	9·52314	31
30	5	9·97435	10·45085		9·54915		9·52350	30
′		Sine.	Tangent.		Cotang.		Cosine.	′

[19 degrees.]   **[70 degrees.]**

[20 degrees.]

′	D.	Cosine.	Cotang.	Diff.	Tangent.	Diff.	Sine.	′
30	5	9.97159	10.42726	38	9.57274	33	9.54433	30
29	5	9.97154	10.42688	38	9.57312	33	9.54466	31
28	5	9.97149	10.42649	39	9.57351	34	9.54500	32
27	4	9.97145	10.42611	38	9.57389	34	9.54534	33
26	5	9.97140	10.42572	39	9.57428	33	9.54567	34
25	5	9.97135	10.42534	38	9.57466	34	9.54601	35
24	5	9.97130	10.42496	38	9.57504	34	9.54635	36
23	4	9.97126	10.42457	38	9.57543	34	9.54668	37
22	5	9.97121	10.42419	39	9.57581	33	9.54702	38
21	5	9.97116	10.42381	38	9.57619	34	9.54735	39
20	5	9.97111	10.42342	38	9.57658	33	9.54769	40
19	5	9.97107	10.42304	38	9.57696	34	9.54802	41
18	5	9.97102	10.42266	38	9.57734	33	9.54836	42
17	5	9.97097	10.42228	38	9.57772	33	9.54869	43
16	5	9.97092	10.42190	38	9.57810	33	9.54903	44
15	5	9.97087	10.42151	39	9.57849	33	9.54936	45
14	4	9.97083	10.42113	38	9.57887	33	9.54969	46
13	5	9.97078	10.42075	38	9.57925	34	9.55003	47
12	5	9.97073	10.42037	38	9.57963	33	9.55036	48
11	5	9.97068	10.41999	38	9.58001	33	9.55069	49
10	5	9.97063	10.41961	38	9.58039	33	9.55102	50
9	4	9.97059	10.41923	38	9.58077	33	9.55136	51
8	5	9.97054	10.41885	38	9.58115	33	9.55169	52
7	5	9.97049	10.41847	38	9.58153	33	9.55202	53
6	5	9.97044	10.41809	38	9.58191	33	9.55235	54
5	5	9.97039	10.41771	38	9.58229	33	9.55268	55
4	4	9.97035	10.41733	37	9.58267	33	9.55301	56
3	5	9.97030	10.41696	38	9.58304	33	9.55334	57
2	5	9.97025	10.41658	38	9.58342	33	9.55367	58
1	5	9.97020	10.41620	38	9.58380	33	9.55400	59
0	5	9.97015	10.41582	38	9.58418	33	9.55433	60
′		Sine.	Tangent.		Cotang.		Cosine.	′

[69 degrees.]

[20 degrees.]

′	Sine.	Diff.	Tangent.	Diff.	Cotang.	Cosine.	D.	′
0	9.53405	35	9.56107	39	10.43893	9.97299	5	60
1	9.53440	35	9.56146	39	10.43854	9.97294	5	59
2	9.53475	34	9.56185	39	10.43815	9.97289	5	58
3	9.53509	35	9.56224	40	10.43776	9.97285	4	57
4	9.53544	34	9.56264	39	10.43736	9.97280	4	56
5	9.53578	35	9.56303	39	10.43697	9.97276	5	55
6	9.53613	34	9.56342	39	10.43658	9.97271	5	54
7	9.53647	35	9.56381	39	10.43619	9.97266	4	53
8	9.53682	34	9.56420	39	10.43580	9.97262	5	52
9	9.53716	35	9.56459	39	10.43541	9.97257	5	51
10	9.53751	34	9.56498	39	10.43502	9.97252	4	50
11	9.53785	34	9.56537	39	10.43463	9.97248	5	49
12	9.53819	35	9.56576	39	10.43424	9.97243	5	48
13	9.53854	34	9.56615	39	10.43385	9.97238	4	47
14	9.53888	34	9.56654	39	10.43346	9.97234	5	46
15	9.53922	35	9.56693	39	10.43307	9.97229	5	45
16	9.53957	34	9.56732	39	10.43268	9.97224	4	44
17	9.53991	34	9.56771	39	10.43229	9.97220	5	43
18	9.54025	34	9.56810	39	10.43190	9.97215	5	42
19	9.54059	34	9.56849	38	10.43151	9.97210	4	41
20	9.54093	34	9.56887	39	10.43113	9.97206	5	40
21	9.54127	34	9.56926	39	10.43074	9.97201	5	39
22	9.54161	34	9.56965	39	10.43035	9.97196	4	38
23	9.54195	34	9.57004	38	10.42996	9.97192	5	37
24	9.54229	34	9.57042	39	10.42958	9.97187	5	36
25	9.54263	34	9.57081	39	10.42919	9.97182	4	35
26	9.54297	34	9.57120	38	10.42880	9.97178	5	34
27	9.54331	34	9.57158	39	10.42842	9.97173	5	33
28	9.54365	34	9.57197	38	10.42803	9.97168	4	32
29	9.54399	34	9.57235	39	10.42765	9.97163	5	31
30	9.54433		9.57274		10.42726	9.97159		30
′			Cotang.		Tangent.	Sine.	D.	′

[69 degrees.]

[21 degrees.]

'	D.	Cosine.	Cotang.	Diff.	Tangent.	Diff.	Sine.	'
30	5	9·96868	10·40460	37	9·59540	32	9·56408	30
29	5	9·96863	10·40423	37	9·59577	32	9·56440	31
28	5	9·96858	10·40386	37	9·59614	32	9·56472	32
27	5	9·96853	10·40349	37	9·59651	32	9·56504	33
26	5	9·96848	10·40312	37	9·59688	32	9·56536	34
25	5	9·96843	10·40275	37	9·59725	31	9·56568	35
24	5	9·96838	10·40238	37	9·59762	32	9·56599	36
23	5	9·96833	10·40201	37	9·59799	32	9·56631	37
22	5	9·96828	10·40165	36	9·59835	32	9·56663	38
21	5	9·96823	10·40128	37	9·59872	32	9·56695	39
20	5	9·96818	10·40091	37	9·59909	31	9·56727	40
19	5	9·96813	10·40054	37	9·59946	32	9·56759	41
18	5	9·96808	10·40017	37	9·59983	32	9·56790	42
17	5	9·96803	10·39981	36	9·60019	32	9·56822	43
16	5	9·96798	10·39944	37	9·60056	31	9·56854	44
15	5	9·96793	10·39907	37	9·60093	32	9·56886	45
14	5	9·96788	10·39870	36	9·60130	32	9·56917	46
13	5	9·96783	10·39834	37	9·60166	31	9·56949	47
12	6	9·96778	10·39797	37	9·60203	32	9·56980	48
11	6	9·96772	10·39760	36	9·60240	32	9·57012	49
10	5	9·96767	10·39724	37	9·60276	31	9·57044	50
9	5	9·96762	10·39687	36	9·60313	32	9·57075	51
8	5	9·96757	10·39651	37	9·60349	31	9·57107	52
7	5	9·96752	10·39614	36	9·60386	31	9·57138	53
6	5	9·96747	10·39578	37	9·60422	32	9·57169	54
5	5	9·96742	10·39541	36	9·60459	31	9·57201	55
4	5	9·96737	10·39505	37	9·60495	31	9·57232	56
3	5	9·96732	10·39468	36	9·60532	32	9·57264	57
2	5	9·96727	10·39432	37	9·60568	31	9·57295	58
1	5	9·96722	10·39395	36	9·60605	31	9·57326	59
0	5	9·96717	10·39359		9·60641	33	9·57358	60
'		Sine.	Tangent.		Cotang.		Cosine.	'

[68 degrees.]

[21 degrees.]

'	Sine.	Diff.	Tangent.	Diff.	Cotang.	Cosine.	D.	'
0	9·55433	33	9·58418	37	10·41582	9·97015	5	60
1	9·55466	33	9·58455	38	10·41545	9·97010	5	59
2	9·55499	33	9·58493	38	10·41507	9·97005	4	58
3	9·55532	32	9·58531	38	10·41469	9·97001	5	57
4	9·55564	33	9·58569	37	10·41431	9·96996	5	56
5	9·55597	33	9·58606	38	10·41394	9·96991	5	55
6	9·55630	33	9·58644	37	10·41356	9·96986	5	54
7	9·55663	32	9·58681	38	10·41319	9·96981	5	53
8	9·55695	33	9·58719	38	10·41281	9·96976	5	52
9	9·55728	33	9·58757	37	10·41243	9·96971	5	51
10	9·55761	32	9·58794	38	10·41206	9·96966	5	50
11	9·55793	33	9·58832	37	10·41168	9·96962	4	49
12	9·55826	32	9·58869	38	10·41131	9·96957	5	48
13	9·55858	33	9·58907	37	10·41093	9·96952	5	47
14	9·55891	32	9·58944	37	10·41056	9·96947	5	46
15	9·55923	33	9·58981	38	10·41019	9·96942	5	45
16	9·55956	32	9·59019	37	10·40981	9·96937	5	44
17	9·55988	33	9·59056	38	10·40944	9·96932	5	43
18	9·56021	32	9·59094	37	10·40906	9·96927	5	42
19	9·56053	32	9·59131	37	10·40869	9·96922	5	41
20	9·56085	33	9·59168	37	10·40832	9·96917	5	40
21	9·56118	32	9·59205	38	10·40795	9·96912	5	39
22	9·56150	32	9·59243	37	10·40757	9·96907	4	38
23	9·56182	33	9·59280	37	10·40720	9·96903	5	37
24	9·56215	32	9·59317	37	10·40683	9·96898	5	36
25	9·56247	32	9·59354	37	10·40646	9·96893	5	35
26	9·56279	32	9·59391	38	10·40609	9·96888	5	34
27	9·56311	32	9·59429	37	10·40571	9·96883	5	33
28	9·56343	32	9·59466	37	10·40534	9·96878	5	32
29	9·56375	33	9·59503	37	10·40497	9·96873	5	31
30	9·56408		9·59540		10·40460	9·96868	5	30
'	Cosine.		Tangent.		Cotang.	Sine.		'

[68 degrees.]

[22 degrees.]   **[67 degrees.]**

′	D.	Cosine.	Cotang.	Diff.	Tangent.	Diff.	Sine.	′
30	6	9·96562	10·38278	36	9·61722	30	9·58284	30
29	5	9·96556	10·38242	36	9·61758	31	9·58314	31
28	5	9·96551	10·38206	36	9·61794	31	9·58345	32
27	5	9·96546	10·38170	35	9·61830	30	9·58375	33
26	6	9·96541	10·38135	36	9·61865	31	9·58406	34
25	5	9·96535	10·38099	35	9·61901	30	9·58436	35
24	5	9·96530	10·38064	36	9·61936	31	9·58467	36
23	5	9·96525	10·38028	36	9·61972	30	9·58497	37
22	5	9·96520	10·37992	35	9·62008	30	9·58527	38
21	6	9·96514	10·37957	36	9·62043	30	9·58557	39
20	5	9·96509	10·37921	35	9·62079	31	9·58588	40
19	5	9·96504	10·37886	36	9·62114	30	9·58618	41
18	5	9·96498	10·37850	35	9·62150	30	9·58648	42
17	5	9·96493	10·37815	36	9·62185	30	9·58678	43
16	6	9·96488	10·37779	35	9·62221	31	9·58709	44
15	5	9·96483	10·37744	36	9·62256	30	9·58739	45
14	5	9·96477	10·37708	35	9·62292	30	9·58769	46
13	6	9·96472	10·37673	35	9·62327	30	9·58799	47
12	5	9·96467	10·37638	36	9·62362	30	9·58829	48
11	6	9·96461	10·37602	35	9·62398	30	9·58859	49
10	5	9·96456	10·37567	35	9·62433	30	9·58889	50
9	5	9·96451	10·37532	36	9·62468	30	9·58919	51
8	6	9·96445	10·37496	35	9·62504	30	9·58949	52
7	5	9·96440	10·37461	35	9·62539	30	9·58979	53
6	5	9·96435	10·37426	35	9·62574	30	9·59009	54
5	6	9·96429	10·37391	36	9·62609	30	9·59039	55
4	5	9·96424	10·37355	35	9·62645	30	9·59069	56
3	5	9·96419	10·37320	35	9·62680	29	9·59098	57
2	6	9·96413	10·37285	35	9·62715	30	9·59128	58
1	5	9·96408	10·37250	35	9·62750	30	9·59158	59
0	5	9·96403	10·37215		9·62785		9·59188	60
′		Sine.	Tangent.		Cotang.		Cosine.	′

[22 degrees.]   **[67 degrees.]**

123

[22 degrees.]   **[67 degrees.]**

′	Sine.	Diff.	Tangent.	Diff.	Cotang.	Cosine.	D.	′
0	9·57358	31	9·60641	36	10·39359	9·96717	6	60
1	9·57389	31	9·60677	37	10·39323	9·96711	5	59
2	9·57420	31	9·60714	36	10·39286	9·96706	5	58
3	9·57451	31	9·60750	36	10·39250	9·96701	5	57
4	9·57482	32	9·60786	37	10·39214	9·96696	5	56
5	9·57514	31	9·60823	36	10·39177	9·96691	5	55
6	9·57545	31	9·60859	36	10·39141	9·96686	5	54
7	9·57576	31	9·60895	36	10·39105	9·96681	5	53
8	9·57607	31	9·60931	36	10·39069	9·96676	6	52
9	9·57638	31	9·60967	37	10·39033	9·96670	5	51
10	9·57669	31	9·61004	36	10·38996	9·96665	5	50
11	9·57700	31	9·61040	36	10·38960	9·96660	5	49
12	9·57731	31	9·61076	36	10·38924	9·96655	5	48
13	9·57762	31	9·61112	36	10·38888	9·96650	5	47
14	9·57793	31	9·61148	36	10·38852	9·96645	6	46
15	9·57824	31	9·61184	36	10·38816	9·96640	5	45
16	9·57855	30	9·61220	36	10·38780	9·96634	5	44
17	9·57885	31	9·61256	36	10·38744	9·96629	5	43
18	9·57916	31	9·61292	36	10·38708	9·96624	5	42
19	9·57947	31	9·61328	36	10·38672	9·96619	6	41
20	9·57978	30	9·61364	36	10·38636	9·96614	5	40
21	9·58008	31	9·61400	36	10·38600	9·96608	5	39
22	9·58039	31	9·61436	36	10·38564	9·96603	5	38
23	9·58070	31	9·61472	36	10·38528	9·96598	5	37
24	9·58101	30	9·61508	36	10·38492	9·96593	6	36
25	9·58131	31	9·61544	35	10·38456	9·96588	5	35
26	9·58162	30	9·61579	36	10·38421	9·96582	5	34
27	9·58192	31	9·61615	36	10·38385	9·96577	5	33
28	9·58223	31	9·61651	36	10·38349	9·96572	6	32
29	9·58253	30	9·61687	35	10·38313	9·96567	5	31
30	9·58284	31	9·61722		10·38278	9·96562	5	30
′			Tangent.		Cotang.	Sine.		′

[22 degrees.]   **[67 degrees.]**

[23 degrees.]   **[66 degrees.]**

′	D.	Cosine.	Cotang.	Diff.	Tangent.	Diff.	Sine.	′
30	6	9·96240	10·36170	35	9·63830	29	9·60070	30
29	5	9·96234	10·36135	34	9·63865	29	9·60099	31
28	6	9·96229	10·36101	35	9·63899	29	9·60128	32
27	5	9·96223	10·36066	34	9·63934	29	9·60157	33
26	6	9·96218	10·36032	35	9·63968	29	9·60186	34
25	5	9·96212	10·35997	34	9·64003	29	9·60215	35
24	6	9·96207	10·35963	35	9·64037	29	9·60244	36
23	5	9·96201	10·35928	34	9·64072	29	9·60273	37
22	6	9·96196	10·35894	34	9·64106	29	9·60302	38
21	5	9·96190	10·35860	35	9·64140	28	9·60331	39
20	6	9·96185	10·35825	34	9·64175	29	9·60359	40
19	5	9·96179	10·35791	34	9·64209	29	9·60388	41
18	6	9·96174	10·35757	35	9·64243	29	9·60417	42
17	6	9·96168	10·35722	34	9·64278	28	9·60446	43
16	5	9·96162	10·35688	34	9·64312	29	9·60474	44
15	6	9·96157	10·35654	35	9·64346	29	9·60503	45
14	5	9·96151	10·35619	34	9·64381	29	9·60532	46
13	6	9·96146	10·35585	34	9·64415	28	9·60561	47
12	5	9·96140	10·35551	34	9·64449	29	9·60589	48
11	6	9·96135	10·35517	34	9·64483	28	9·60618	49
10	6	9·96129	10·35483	35	9·64517	29	9·60646	50
9	5	9·96123	10·35448	34	9·64552	29	9·60675	51
8	6	9·96118	10·35414	34	9·64586	28	9·60704	52
7	5	9·96112	10·35380	34	9·64620	29	9·60732	53
6	6	9·96107	10·35346	34	9·64654	28	9·60761	54
5	6	9·96101	10·35312	34	9·64688	29	9·60789	55
4	5	9·96095	10·35278	34	9·64722	28	9·60818	56
3	6	9·96090	10·35244	34	9·64756	29	9·60846	57
2	5	9·96084	10·35210	34	9·64790	28	9·60875	58
1	6	9·96079	10·35176	34	9·64824	28	9·60903	59
0	6	9·96073	10·35142	34	9·64858	28	9·60931	60
′		Sine.	Tangent.		Cotang.		Cosine.	′

[66 degrees.]

124

[23 degrees.]   **[66 degrees.]**

′	Sine.	Diff.	Tangent.	Diff.	Cotang.	Cosine.	D.	′
0	9·59188	30	9·62785	35	10·37215	9·96403	6	60
1	9·59218	29	9·62820	35	10·37180	9·96397	5	59
2	9·59247	30	9·62855	35	10·37145	9·96392	5	58
3	9·59277	30	9·62890	36	10·37110	9·96387	6	57
4	9·59307	29	9·62926	35	10·37074	9·96381	5	56
5	9·59336	30	9·62961	35	10·37039	9·96376	6	55
6	9·59366	30	9·62996	35	10·37004	9·96370	5	54
7	9·59396	29	9·63031	35	10·36969	9·96365	5	53
8	9·59425	30	9·63066	35	10·36934	9·96360	6	52
9	9·59455	29	9·63101	34	10·36899	9·96354	5	51
10	9·59484	30	9·63135	35	10·36865	9·96349	6	50
11	9·59514	29	9·63170	35	10·36830	9·96343	5	49
12	9·59543	30	9·63205	35	10·36795	9·96338	5	48
13	9·59573	29	9·63240	35	10·36760	9·96333	6	47
14	9·59602	30	9·63275	35	10·36725	9·96327	5	46
15	9·59632	29	9·63310	35	10·36690	9·96322	6	45
16	9·59661	29	9·63345	34	10·36655	9·96316	5	44
17	9·59690	30	9·63379	35	10·36621	9·96311	6	43
18	9·59720	29	9·63414	35	10·36586	9·96305	5	42
19	9·59749	29	9·63449	35	10·36551	9·96300	6	41
20	9·59778	30	9·63484	35	10·36516	9·96294	5	40
21	9·59808	29	9·63519	34	10·36481	9·96289	5	39
22	9·59837	29	9·63553	35	10·36447	9·96284	6	38
23	9·59866	29	9·63588	35	10·36412	9·96278	5	37
24	9·59895	29	9·63623	34	10·36377	9·96273	6	36
25	9·59924	30	9·63657	35	10·36343	9·96267	5	35
26	9·59954	29	9·63692	34	10·36308	9·96262	6	34
27	9·59983	29	9·63726	35	10·36274	9·96256	5	33
28	9·60012	29	9·63761	35	10·36239	9·962·1	6	32
29	9·60041	29	9·63796	34	10·36204	9·9624·	5	31
30	9·60070		9·63830		10·36170	9·96240		30
′			Cotang.			Tangent.	Sine.	′

[66 degrees.]

[24 degrees.] — **[65 degrees.]**

′	D.	Cosine.	Cotang.	Diff.	Tangent.	Diff.	Sine.	′
30	5	9.95902	10.34130	34	9.65870	27	9.61773	30
29	6	9.95897	10.34096	33	9.65904	28	9.61800	31
28	6	9.95891	10.34063	34	9.65937	28	9.61828	32
27	6	9.95885	10.34029	33	9.65971	27	9.61856	33
26	6	9.95879	10.33996	34	9.66004	28	9.61883	34
25	6	9.95873	10.33962	33	9.66038	28	9.61911	35
24	5	9.95868	10.33929	33	9.66071	27	9.61939	36
23	6	9.95862	10.33896	34	9.66104	28	9.61966	37
22	6	9.95856	10.33862	33	9.66138	27	9.61994	38
21	6	9.95850	10.33829	33	9.66171	28	9.62021	39
20	6	9.95844	10.33796	34	9.66204	27	9.62049	40
19	5	9.95839	10.33762	33	9.66238	28	9.62076	41
18	6	9.95833	10.33729	33	9.66271	27	9.62104	42
17	6	9.95827	10.33696	33	9.66304	28	9.62131	43
16	6	9.95821	10.33663	34	9.66337	27	9.62159	44
15	6	9.95815	10.33629	33	9.66371	27	9.62186	45
14	5	9.95810	10.33596	33	9.66404	28	9.62214	46
13	6	9.95804	10.33563	33	9.66437	27	9.62241	47
12	6	9.95798	10.33530	33	9.66470	27	9.62268	48
11	6	9.95792	10.33497	34	9.66503	28	9.62296	49
10	6	9.95786	10.33463	33	9.66537	27	9.62323	50
9	5	9.95780	10.33430	33	9.66570	27	9.62350	51
8	6	9.95775	10.33397	33	9.66603	27	9.62377	52
7	6	9.95769	10.33364	33	9.66636	28	9.62405	53
6	6	9.95763	10.33331	33	9.66669	27	9.62432	54
5	6	9.95757	10.33298	33	9.66702	27	9.62459	55
4	6	9.95751	10.33265	33	9.66735	27	9.62486	56
3	6	9.95745	10.33232	33	9.66768	27	9.62513	57
2	6	9.95739	10.33199	33	9.66801	28	9.62541	58
1	6	9.95733	10.33166	33	9.66834	27	9.62568	59
0	5	9.95728	10.33133		9.66867	27	9.62595	60
′		Sine.	Tangent.		Cotang.		Cosine.	′

[65 degrees.]

125

[24 degrees.] — **[65 degrees.]**

′	Sine.	Diff.	Tangent.	Diff.	Cotang.	Cosine.	D.	′
0	9.60931	29	9.64858	34	10.35142	9.96073	6	60
1	9.60960	28	9.64892	34	10.35108	9.96067	5	59
2	9.60988	28	9.64926	34	10.35074	9.96062	6	58
3	9.61016	29	9.64960	34	10.35040	9.96056	6	57
4	9.61045	28	9.64994	34	10.35006	9.96050	5	56
5	9.61073	28	9.65028	34	10.34972	9.96045	6	55
6	9.61101	28	9.65062	34	10.34938	9.96039	5	54
7	9.61129	29	9.65096	34	10.34904	9.96034	6	53
8	9.61158	28	9.65130	34	10.34870	9.96028	6	52
9	9.61186	28	9.65164	33	10.34836	9.96022	5	51
10	9.61214	28	9.65197	34	10.34803	9.96017	6	50
11	9.61242	28	9.65231	34	10.34769	9.96011	6	49
12	9.61270	28	9.65265	34	10.34735	9.96005	5	48
13	9.61298	28	9.65299	34	10.34701	9.96000	6	47
14	9.61326	28	9.65333	33	10.34667	9.95994	6	46
15	9.61354	28	9.65366	34	10.34634	9.95988	6	45
16	9.61382	29	9.65400	34	10.34600	9.95982	5	44
17	9.61411	27	9.65434	33	10.34566	9.95977	6	43
18	9.61438	28	9.65467	34	10.34533	9.95971	6	42
19	9.61466	28	9.65501	34	10.34499	9.95965	5	41
20	9.61494	28	9.65535	33	10.34465	9.95960	6	40
21	9.61522	28	9.65568	34	10.34432	9.95954	6	39
22	9.61550	28	9.65602	34	10.34398	9.95948	6	38
23	9.61578	28	9.65636	33	10.34364	9.95942	5	37
24	9.61606	28	9.65669	34	10.34331	9.95937	6	36
25	9.61634	28	9.65703	33	10.34297	9.95931	6	35
26	9.61662	27	9.65736	34	10.34264	9.95925	5	34
27	9.61689	28	9.65770	33	10.34230	9.95920	6	33
28	9.61717	28	9.65803	34	10.34197	9.95914	6	32
29	9.61745	28	9.65837	33	10.34163	9.95908	6	31
30	9.61773		9.65870	33	10.34130	9.95902		30
′	Cosine.		Cotang.		Tangent.	Sine.		′

[65 degrees.]

[25 degrees.]   [64¾ degrees.]

'	D.	Cosine.	Cotang.	Diff	Tangent.	Diff	Sine.	'
30	6	9·95549	10·32150	32	9·67850	27	9·63398	30
31	6	9·95543	10·32118	33	9·67882	26	9·63425	29
32	6	9·95537	10·32085	32	9·67915	27	9·63451	28
33	6	9·95531	10·32053	33	9·67947	26	9·63478	27
34	6	9·95525	10·32020	32	9·67980	27	9·63504	26
35	6	9·95519	10·31988	32	9·68012	26	9·63531	25
36	6	9·95513	10·31956	33	9·68044	26	9·63557	24
37	6	9·95507	10·31923	32	9·68077	27	9·63583	23
38	7	9·95500	10·31891	33	9·68109	26	9·63610	22
39	6	9·95494	10·31858	32	9·68142	26	9·63636	21
40	6	9·95488	10·31826	32	9·68174	27	9·63662	20
41	6	9·95482	10·31794	33	9·68206	26	9·63689	19
42	6	9·95476	10·31761	32	9·68239	26	9·63715	18
43	6	9·95470	10·31729	32	9·68271	26	9·63741	17
44	6	9·95464	10·31697	33	9·68303	27	9·63767	16
45	6	9·95458	10·31664	32	9·68336	26	9·63794	15
46	6	9·95452	10·31632	32	9·68368	26	9·63820	14
47	6	9·95446	10·31600	32	9·68400	26	9·63846	13
48	6	9·95440	10·31568	33	9·68432	26	9·63872	12
49	7	9·95434	10·31535	32	9·68465	26	9·63898	11
50	6	9·95427	10·31503	32	9·68497	26	9·63924	10
51	6	9·95421	10·31471	32	9·68529	26	9·63950	9
52	6	9·95415	10·31439	32	9·68561	26	9·63976	8
53	6	9·95409	10·31407	33	9·68593	26	9·64002	7
54	6	9·95403	10·31374	32	9·68626	26	9·64028	6
55	6	9·95397	10·31342	32	9·68658	26	9·64054	5
56	6	9·95391	10·31310	32	9·68690	26	9·64080	4
57	6	9·95384	10·31278	32	9·68722	26	9·64106	3
58	6	9·95378	10·31246	32	9·68754	26	9·64132	2
59	7	9·95372	10·31214	32	9·68786	26	9·64158	1
60	6	9·95366	10·31182		9·68818		9·64184	0
		Sine.	Tangent.		Cotang.		Cosine.	'

[25 degrees.]   [64 degrees.]

'	Sine.	Diff	Tangent.	Diff	Cotang.	Cosine.	D.	'
0	9·62595	27	9·66867	33	10·33133	9·95728	6	60
1	9·62622	27	9·66900	33	10·33100	9·95722	6	59
2	9·62649	27	9·66933	33	10·33067	9·95716	6	58
3	9·62676	27	9·66966	33	10·33034	9·95710	6	57
4	9·62703	27	9·66999	33	10·33001	9·95704	6	56
5	9·62730	27	9·67032	33	10·32968	9·95698	6	55
6	9·62757	27	9·67065	33	10·32935	9·95692	6	54
7	9·62784	27	9·67098	33	10·32902	9·95686	6	53
8	9·62811	27	9·67131	32	10·32869	9·95680	6	52
9	9·62838	27	9·67163	33	10·32837	9·95674	6	51
10	9·62865	27	9·67196	33	10·32804	9·95668	6	50
11	9·62892	26	9·67229	33	10·32771	9·95663	6	49
12	9·62918	27	9·67262	33	10·32738	9·95657	6	48
13	9·62945	27	9·67295	32	10·32705	9·95651	6	47
14	9·62972	27	9·67327	33	10·32673	9·95645	6	46
15	9·62999	27	9·67360	33	10·32640	9·95639	6	45
16	9·63026	26	9·67393	33	10·32607	9·95633	6	44
17	9·63052	27	9·67426	32	10·32574	9·95627	6	43
18	9·63079	27	9·67458	33	10·32542	9·95621	6	42
19	9·63106	27	9·67491	33	10·32509	9·95615	6	41
20	9·63133	26	9·67524	32	10·32476	9·95609	6	40
21	9·63159	27	9·67556	33	10·32444	9·95603	6	39
22	9·63186	27	9·67589	33	10·32411	9·95597	6	38
23	9·63213	26	9·67622	32	10·32378	9·95591	6	37
24	9·63239	27	9·67654	33	10·32346	9·95585	6	36
25	9·63266	26	9·67687	32	10·32313	9·95579	6	35
26	9·63292	27	9·67719	33	10·32281	9·95573	6	34
27	9·63319	26	9·67752	33	10·32248	9·95567	6	33
28	9·63345	27	9·67785	32	10·32215	9·95561	6	32
29	9·63372	26	9·67817	33	10·32183	9·95555	6	31
30	9·63398		9·67850		10·32150	9·95549	6	30
	Cosine.		Cotang.		Tangent.	Sine.		'

[26 degrees.]   [63 degrees.]

′	D.	Cosine.	Cotang.	Diff.	Tangent.	Diff.	Sine.	′
30	6	9·95179	10·30226	31	9·69774	25	9·64953	30
29	6	9·95173	10·30195	32	9·69805	25	9·64978	31
28	6	9·95167	10·30163	31	9·69837	26	9·65003	32
27	7	9·95160	10·30132	32	9·69868	25	9·65029	33
26	6	9·95154	10·30100	32	9·69900	25	9·65054	34
25	6	9·95148	10·30068	31	9·69932	25	9·65079	35
24	7	9·95141	10·30037	32	9·69963	26	9·65104	36
23	6	9·95135	10·30005	31	9·69995	25	9·65130	37
22	6	9·95129	10·29974	32	9·70026	25	9·65155	38
21	7	9·95122	10·29942	31	9·70058	25	9·65180	39
20	6	9·95116	10·29911	32	9·70089	25	9·65205	40
19	6	9·95110	10·29879	31	9·70121	25	9·65230	41
18	7	9·95103	10·29848	32	9·70152	26	9·65255	42
17	7	9·95097	10·29816	32	9·70184	25	9·65281	43
16	6	9·95090	10·29785	32	9·70215	25	9·65306	44
15	6	9·95084	10·29753	31	9·70247	25	9·65331	45
14	7	9·95078	10·29722	31	9·70278	25	9·65356	46
13	6	9·95071	10·29691	32	9·70309	25	9·65381	47
12	6	9·95065	10·29659	31	9·70341	25	9·65406	48
11	7	9·95059	10·29628	32	9·70372	25	9·65431	49
10	6	9·95052	10·29596	31	9·70404	25	9·65456	50
9	7	9·95046	10·29565	31	9·70435	25	9·65481	51
8	6	9·95039	10·29534	31	9·70466	25	9·65506	52
7	7	9·95033	10·29502	31	9·70498	25	9·65531	53
6	7	9·95027	10·29471	31	9·70529	24	9·65556	54
5	6	9·95020	10·29440	31	9·70560	25	9·65580	55
4	7	9·95014	10·29408	31	9·70592	25	9·65605	56
3	6	9·95007	10·29377	31	9·70623	25	9·65630	57
2	6	9·95001	10·29346	31	9·70654	25	9·65655	58
1	7	9·94995	10·29315	32	9·70685	25	9·65680	59
0	7	9·94988	10·29283		9·70717		9·65705	60
′		Sine.	Tangent.		Cotang.		Cosine.	′

[26 degrees.]   [63 degrees.]

127

[26 degrees.]   [63 degrees.]

′	D.	Cosine.	Cotang.	Diff.	Tangent.	Diff.	Sine.	′
0	6	9·95366	10·31182	32	9·68818	26	9·64184	60
1	6	9·95360	10·31150	32	9·68850	26	9·64210	59
2	6	9·95354	10·31118	32	9·68882	26	9·64236	58
3	6	9·95348	10·31086	32	9·68914	26	9·64262	57
4	6	9·95341	10·31054	32	9·68946	26	9·64288	56
5	7	9·95335	10·31022	32	9·68978	25	9·64313	55
6	6	9·95329	10·30990	32	9·69010	26	9·64339	54
7	6	9·95323	10·30958	32	9·69042	26	9·64365	53
8	6	9·95317	10·30926	32	9·69074	26	9·64391	52
9	6	9·95310	10·30894	32	9·69106	26	9·64417	51
10	6	9·95304	10·30862	32	9·69138	25	9·64442	50
11	7	9·95298	10·30830	32	9·69170	26	9·64468	49
12	6	9·95292	10·30798	32	9·69202	26	9·64494	48
13	6	9·95286	10·30766	32	9·69234	25	9·64519	47
14	7	9·95279	10·30734	32	9·69266	26	9·64545	46
15	6	9·95273	10·30702	31	9·69298	26	9·64571	45
16	6	9·95267	10·30671	32	9·69329	25	9·64596	44
17	7	9·95261	10·30639	32	9·69361	26	9·64622	43
18	6	9·95254	10·30607	32	9·69393	25	9·64647	42
19	6	9·95248	10·30575	32	9·69425	26	9·64673	41
20	7	9·95242	10·30543	31	9·69457	25	9·64698	40
21	6	9·95236	10·30512	32	9·69488	26	9·64724	39
22	6	9·95229	10·30480	32	9·69520	25	9·64749	38
23	7	9·95223	10·30448	32	9·69552	26	9·64775	37
24	6	9·95217	10·30416	32	9·69584	25	9·64800	36
25	6	9·95210	10·30385	31	9·69615	26	9·64826	35
26	7	9·95204	10·30353	32	9·69647	25	9·64851	34
27	6	9·95198	10·30321	32	9·69679	26	9·64877	33
28	6	9·95192	10·30290	31	9·69710	25	9·64902	32
29	7	9·95185	10·30258	32	9·69742	25	9·64927	31
30	6	9·95179	10·30226		9·69774		9·64953	30
′		Sine.	Tangent.		Cotang.		Cosine.	′

[26 degrees.]   [63 degrees.]

[27 degrees.]   **[62 degrees.]**

'	D.	Cosine	Cotang.	Diff.	Tangent	Diff.	Sine	Diff.	'
30	7	9·94793	10·28352	31	9·71648	31	9·66441	24	30
29	6	9·94786	10·28321	31	9·71679	30	9·66465	24	31
28	7	9·94780	10·28291	31	9·71709	31	9·66489	24	32
27	6	9·94773	10·28260	31	9·71740	31	9·66513	24	33
26	7	9·94767	10·28229	31	9·71771	31	9·66537	25	34
25	7	9·94760	10·28198	31	9·71802	31	9·66562	24	35
24	6	9·94753	10·28167	30	9·71833	30	9·66586	24	36
23	7	9·94747	10·28137	31	9·71863	31	9·66610	24	37
22	7	9·94740	10·28106	31	9·71894	31	9·66634	24	38
21	7	9·94734	10·28075	31	9·71925	30	9·66658	24	39
20	7	9·94727	10·28045	30	9·71955	31	9·66682	24	40
19	6	9·94720	10·28014	31	9·71986	31	9·66706	25	41
18	7	9·94714	10·27983	31	9·72017	31	9·66731	24	42
17	7	9·94707	10·27952	30	9·72048	30	9·66755	24	43
16	6	9·94700	10·27922	31	9·72078	31	9·66779	24	44
15	7	9·94694	10·27891	31	9·72109	31	9·66803	24	45
14	7	9·94687	10·27860	30	9·72140	30	9·66827	24	46
13	6	9·94680	10·27830	31	9·72170	31	9·66851	24	47
12	7	9·94674	10·27799	30	9·72201	30	9·66875	24	48
11	7	9·94667	10·27769	31	9·72231	31	9·66899	23	49
10	6	9·94660	10·27738	31	9·72262	31	9·66922	24	50
9	7	9·94654	10·27707	30	9·72293	30	9·66946	24	51
8	7	9·94647	10·27677	31	9·72323	31	9·66970	24	52
7	6	9·94640	10·27646	30	9·72354	30	9·66994	24	53
6	7	9·94634	10·27616	31	9·72384	31	9·67018	24	54
5	7	9·94627	10·27585	30	9·72415	30	9·67042	24	55
4	6	9·94620	10·27555	31	9·72445	31	9·67066	24	56
3	7	9·94614	10·27524	30	9·72476	30	9·67090	23	57
2	7	9·94607	10·27494	31	9·72506	31	9·67113	24	58
1	7	9·94600	10·27463	30	9·72537	30	9·67137	24	59
0	7	9·94593	10·27433		9·72567		9·67161		60
'		Sine.	Tangent.		Cotang.		Cosine.		'

128

[27 degrees.]   **[62 degrees.]**

'	Sine.	Diff.	Tangent.	Diff.	Cotang.	Cosine.	D.	'
0	9·65705	24	9·70717	31	10·29283	9·94988	6	60
1	9·65729	25	9·70748	31	10·29252	9·94982	6	59
2	9·65754	25	9·70779	31	10·29221	9·94975	7	58
3	9·65779	25	9·70810	31	10·29190	9·94969	6	57
4	9·65804	24	9·70841	32	10·29159	9·94962	7	56
5	9·65828	25	9·70873	31	10·29127	9·94956	7	55
6	9·65853	25	9·70904	31	10·29096	9·94949	6	54
7	9·65878	24	9·70935	31	10·29065	9·94943	7	53
8	9·65902	25	9·70966	31	10·29034	9·94936	6	52
9	9·65927	25	9·70997	31	10·29003	9·94930	7	51
10	9·65952	24	9·71028	31	10·28972	9·94923	6	50
11	9·65976	25	9·71059	31	10·28941	9·94917	7	49
12	9·66001	24	9·71090	31	10·28910	9·94911	7	48
13	9·66025	25	9·71121	32	10·28879	9·94904	6	47
14	9·66050	25	9·71153	31	10·28847	9·94898	7	46
15	9·66075	24	9·71184	31	10·28816	9·94891	6	45
16	9·66099	25	9·71215	31	10·28785	9·94885	7	44
17	9·66124	24	9·71246	31	10·28754	9·94878	7	43
18	9·66148	25	9·71277	31	10·28723	9·94871	6	42
19	9·66173	24	9·71308	31	10·28692	9·94865	7	41
20	9·66197	24	9·71339	31	10·28661	9·94858	6	40
21	9·66221	25	9·71370	31	10·28630	9·94852	7	39
22	9·66246	24	9·71401	30	10·28599	9·94845	6	38
23	9·66270	25	9·71431	31	10·28569	9·94839	7	37
24	9·66295	24	9·71462	31	10·28538	9·94832	6	36
25	9·66319	24	9·71493	31	10·28507	9·94826	7	35
26	9·66343	25	9·71524	31	10·28476	9·94819	6	34
27	9·66368	24	9·71555	31	10·28445	9·94813	7	33
28	9·66392	24	9·71586	31	10·28414	9·94806	7	32
29	9·66416	25	9·71617	31	10·28383	9·94799	6	31
30	9·66441		9·71648		10·28352	9·94793		30
'	Cosine.		Cotang.		Tangent.	Sine.		'

[28 degrees.]

′	D.	Cosine.	Cotang.	Diff.	Tangent.	Diff.	Sine.	′
30	7	9·94390	10·26524	31	9·73476	24	9·67866	30
29	7	9·94383	10·26493	30	9·73507	23	9·67890	31
28	7	9·94376	10·26463	30	9·73537	23	9·67913	32
27	7	9·94369	10·26433	30	9·73567	23	9·67936	33
26	7	9·94362	10·26403	30	9·73597	23	9·67959	34
25	6	9·94355	10·26373	30	9·73627	23	9·67982	35
24	7	9·94349	10·26343	30	9·73657	24	9·68006	36
23	7	9·94342	10·26313	30	9·73687	23	9·68029	37
22	7	9·94335	10·26283	30	9·73717	23	9·68052	38
21	7	9·94328	10·26253	30	9·73747	23	9·68075	39
20	7	9·94321	10·26223	30	9·73777	23	9·68098	40
19	7	9·94314	10·26193	30	9·73807	23	9·68121	41
18	7	9·94307	10·26163	30	9·73837	23	9·68144	42
17	7	9·94300	10·26133	30	9·73867	23	9·68167	43
16	7	9·94293	10·26103	30	9·73897	23	9·68190	44
15	7	9·94286	10·26073	30	9·73927	24	9·68213	45
14	6	9·94279	10·26043	30	9·73957	23	9·68237	46
13	7	9·94273	10·26013	30	9·73987	23	9·68260	47
12	7	9·94266	10·25983	30	9·74017	22	9·68283	48
11	7	9·94259	10·25953	30	9·74047	23	9·68305	49
10	7	9·94252	10·25923	30	9·74077	23	9·68328	50
9	7	9·94245	10·25893	30	9·74107	23	9·68351	51
8	7	9·94238	10·25863	30	9·74137	23	9·68374	52
7	7	9·94231	10·25834	29	9·74166	23	9·68397	53
6	7	9·94224	10·25804	30	9·74196	23	9·68420	54
5	7	9·94217	10·25774	30	9·74226	23	9·68443	55
4	7	9·94210	10·25744	30	9·74256	23	9·68466	56
3	7	9·94203	10·25714	30	9·74286	23	9·68489	57
2	7	9·94196	10·25684	29	9·74316	22	9·68512	58
1	7	9·94189	10·25655	30	9·74345	23	9·68534	59
0	7	9·94182	10·25625		9·74375		9·68557	60
′		Sine.	Tangent.		Cotang.		Cosine.	′

[61 degrees.]

[28 degrees.]

′	Sine.	Diff.	Tangent.	Diff.	Cotang.	D.	Cosine.	′
0	9·67161	24	9·72567	31	10·27433	6	9·94593	60
1	9·67185	23	9·72598	30	10·27402	7	9·94587	59
2	9·67208	24	9·72628	31	10·27372	7	9·94580	58
3	9·67232	24	9·72659	30	10·27341	6	9·94573	57
4	9·67256	24	9·72689	31	10·27311	7	9·94567	56
5	9·67280	23	9·72720	30	10·27280	7	9·94560	55
6	9·67303	24	9·72750	30	10·27250	7	9·94553	54
7	9·67327	23	9·72780	31	10·27220	7	9·94546	53
8	9·67350	24	9·72811	30	10·27189	6	9·94540	52
9	9·67374	24	9·72841	31	10·27159	7	9·94533	51
10	9·67398	23	9·72872	30	10·27128	7	9·94526	50
11	9·67421	24	9·72902	31	10·27098	7	9·94519	49
12	9·67445	23	9·72932	31	10·27068	7	9·94513	48
13	9·67468	24	9·72963	30	10·27037	7	9·94506	47
14	9·67492	23	9·72993	31	10·27007	6	9·94499	46
15	9·67515	24	9·73023	31	10·26977	7	9·94492	45
16	9·67539	23	9·73054	30	10·26946	7	9·94485	44
17	9·67562	24	9·73084	30	10·26916	7	9·94479	43
18	9·67586	23	9·73114	31	10·26886	7	9·94472	42
19	9·67609	24	9·73145	30	10·26856	7	9·94465	41
20	9·67633	23	9·73175	30	10·26825	7	9·94458	40
21	9·67656	24	9·73205	30	10·26795	7	9·94451	39
22	9·67680	23	9·73235	30	10·26765	7	9·94445	38
23	9·67703	23	9·73265	31	10·26735	7	9·94438	37
24	9·67726	24	9·73295	31	10·26705	7	9·94431	36
25	9·67750	23	9·73326	30	10·26674	7	9·94424	35
26	9·67773	23	9·73356	30	10·26644	7	9·94417	34
27	9·67796	24	9·73386	30	10·26614	7	9·94410	33
28	9·67820	23	9·73416	30	10·26584	7	9·94404	32
29	9·67843	23	9·73446	30	10·26554	7	9·94397	31
30	9·67866		9·73476		10·26524	7	9·94390	30
′	Cosine.		Cotang.		Tangent.		Sine.	′

[61 degrees.]

[29 degrees.]   **[60 degrees.]**

,	D.	Cosine.	Cotang.	Diff.	Tangent.	Diff.	Sine.	,
30	7	9·93970	10·24736	30	9·75264	22	9·69234	30
29	8	9·93963	10·24706	29	9·75294	23	9·69256	31
28	7	9·93955	10·24677	30	9·75323	22	9·69279	32
27	7	9·93948	10·24647	29	9·75353	22	9·69301	33
26	7	9·93941	10·24618	29	9·75382	22	9·69323	34
25	7	9·93934	10·24589	30	9·75411	23	9·69345	35
24	7	9·93927	10·24559	29	9·75441	22	9·69368	36
23	8	9·93920	10·24530	30	9·75470	22	9·69390	37
22	7	9·93912	10·24500	29	9·75500	22	9·69412	38
21	7	9·93905	10·24471	29	9·75529	22	9·69434	39
20	7	9·93898	10·24442	30	9·75558	23	9·69456	40
19	7	9·93891	10·24412	29	9·75588	22	9·69479	41
18	8	9·93884	10·24383	30	9·75617	22	9·69501	42
17	7	9·93876	10·24353	29	9·75647	22	9·69523	43
16	7	9·93869	10·24324	29	9·75676	22	9·69545	44
15	7	9·93862	10·24295	30	9·75705	22	9·69567	45
14	8	9·93855	10·24265	29	9·75735	22	9·69589	46
13	7	9·93847	10·24236	29	9·75764	22	9·69611	47
12	7	9·93840	10·24207	30	9·75793	22	9·69633	48
11	8	9·93833	10·24178	30	9·75822	22	9·69655	49
10	7	9·93826	10·24148	29	9·75852	22	9·69677	50
9	8	9·93819	10·24119	29	9·75881	22	9·69699	51
8	8	9·93811	10·24090	30	9·75910	22	9·69721	52
7	7	9·93804	10·24061	29	9·75939	22	9·69743	53
6	7	9·93797	10·24031	29	9·75969	22	9·69765	54
5	8	9·93789	10·24002	29	9·75998	22	9·69787	55
4	7	9·93782	10·23973	30	9·76027	22	9·69809	56
3	8	9·93775	10·23944	30	9·76056	22	9·69831	57
2	8	9·93768	10·23914	29	9·76086	22	9·69853	58
1	7	9·93760	10·23885	29	9·76115	22	9·69875	59
0	7	9·93753	10·23856		9·76144		9·69897	60
,		Sine.	Tangent.		Cotang.		Cosine.	,

[60 degrees.]

130

[29 degrees.]   **[60 degrees.]**

,	D.	Cosine.	Cotang.	Diff.	Tangent.	Diff.	Sine.	,
0	7	9·94182	10·25625	30	9·74375	23	9·68557	60
1	7	9·94175	10·25595	30	9·74405	23	9·68580	59
2	7	9·94168	10·25565	30	9·74435	22	9·68603	58
3	7	9·94161	10·25535	30	9·74465	23	9·68625	57
4	7	9·94154	10·25506	30	9·74494	23	9·68648	56
5	7	9·94147	10·25476	30	9·74524	23	9·68671	55
6	7	9·94140	10·25446	29	9·74554	22	9·68694	54
7	7	9·94133	10·25417	30	9·74583	23	9·68716	53
8	7	9·94126	10·25387	30	9·74613	23	9·68739	52
9	7	9·94119	10·25357	30	9·74643	22	9·68762	51
10	7	9·94112	10·25327	29	9·74673	23	9·68784	50
11	7	9·94105	10·25298	30	9·74702	23	9·68807	49
12	7	9·94098	10·25268	30	9·74732	23	9·68829	48
13	8	9·94090	10·25238	29	9·74762	23	9·68852	47
14	7	9·94083	10·25209	30	9·74791	22	9·68875	46
15	7	9·94076	10·25179	30	9·74821	23	9·68897	45
16	7	9·94069	10·25149	29	9·74851	23	9·68920	44
17	7	9·94062	10·25120	30	9·74880	22	9·68942	43
18	7	9·94055	10·25090	30	9·74910	23	9·68965	42
19	7	9·94048	10·25061	30	9·74939	22	9·68987	41
20	7	9·94041	10·25031	29	9·74969	23	9·69010	40
21	7	9·94034	10·25002	30	9·74998	23	9·69032	39
22	7	9·94027	10·24972	30	9·75028	23	9·69055	38
23	7	9·94020	10·24942	29	9·75058	22	9·69077	37
24	7	9·94012	10·24913	30	9·75087	23	9·69100	36
25	8	9·94005	10·24883	29	9·75117	22	9·69122	35
26	7	9·93998	10·24854	30	9·75146	22	9·69144	34
27	7	9·93991	10·24824	29	9·75176	23	9·69167	33
28	7	9·93984	10·24795	30	9·75205	22	9·69189	32
29	7	9·93977	10·24765	29	9·75235	23	9·69212	31
30	7	9·93970	10·24736		9·75264	22	9·69234	30
,		Sine.	Tangent.		Cotang.		Cosine.	,

[60 degrees.]

[30 degrees.]   **[59 degrees.]**

′	D.	Cosine.	Cotang.	Diff.	Tangent.	Diff.	Sine.	′
30	7	9·93532	10·22985	29	9·77015	21	9·70547	30
31	8	9·93525	10·22956	29	9·77044	22	9·70568	29
32	7	9·93517	10·22927	28	9·77073	21	9·70590	28
33	8	9·93510	10·22899	29	9·77101	22	9·70611	27
34	8	9·93502	10·22870	29	9·77130	21	9·70633	26
35	8	9·93495	10·22841	29	9·77159	21	9·70654	25
36	7	9·93487	10·22812	29	9·77188	22	9·70675	24
37	8	9·93480	10·22783	29	9·77217	21	9·70697	23
38	7	9·93472	10·22754	28	9·77246	21	9·70718	22
39	8	9·93465	10·22726	29	9·77274	21	9·70739	21
40	7	9·93457	10·22697	29	9·77303	22	9·70761	20
41	8	9·93450	10·22668	28	9·77332	21	9·70782	19
42	7	9·93442	10·22639	29	9·77361	21	9·70803	18
43	8	9·93435	10·22610	28	9·77390	21	9·70824	17
44	8	9·93427	10·22582	29	9·77418	22	9·70846	16
45	7	9·93420	10·22553	29	9·77447	21	9·70867	15
46	8	9·93412	10·22524	29	9·77476	21	9·70888	14
47	7	9·93405	10·22495	28	9·77505	21	9·70909	13
48	8	9·93397	10·22467	29	9·77533	22	9·70931	12
49	7	9·93390	10·22438	29	9·77562	21	9·70952	11
50	8	9·93382	10·22409	28	9·77591	21	9·70973	10
51	8	9·93375	10·22381	29	9·77619	21	9·70994	9
52	7	9·93367	10·22352	29	9·77648	21	9·71015	8
53	8	9·93360	10·22323	28	9·77677	21	9·71036	7
54	8	9·93352	10·22294	28	9·77706	22	9·71058	6
55	7	9·93344	10·22266	29	9·77734	21	9·71079	5
56	8	9·93337	10·22237	28	9·77763	21	9·71100	4
57	8	9·93329	10·22209	29	9·77791	21	9·71121	3
58	7	9·93322	10·22180	29	9·77820	21	9·71142	2
59	8	9·93314	10·22151	28	9·77849	21	9·71163	1
60	7	9·93307	10·22123		9·77877		9·71184	0
′		Sine.	Tangent.		Cotang.		Cosine.	′

[59 degrees.]

131

[30 degrees.]   **[59 degrees.]**

′	D.	Cosine.	Cotang.	Diff.	Tangent.	Diff.	Sine.	′
0	7	9·93753	10·23856	29	9·76144	22	9·69897	60
1	8	9·93746	10·23827	29	9·76173	22	9·69919	59
2	7	9·93738	10·23798	29	9·76202	22	9·69941	58
3	7	9·93731	10·23769	30	9·76231	21	9·69963	57
4	7	9·93724	10·23739	29	9·76261	23	9·69984	56
5	8	9·93717	10·23710	29	9·76290	22	9·70006	55
6	7	9·93709	10·23681	29	9·76319	22	9·70028	54
7	7	9·93702	10·23652	29	9·76348	22	9·70050	53
8	8	9·93695	10·23623	29	9·76377	22	9·70072	52
9	7	9·93687	10·23594	29	9·76406	21	9·70093	51
10	7	9·93680	10·23565	29	9·76435	22	9·70115	50
11	8	9·93673	10·23536	29	9·76464	22	9·70137	49
12	7	9·93665	10·23507	29	9·76493	22	9·70159	48
13	7	9·93658	10·23478	29	9·76522	21	9·70180	47
14	8	9·93650	10·23449	29	9·76551	22	9·70202	46
15	7	9·93643	10·23420	29	9·76580	22	9·70224	45
16	7	9·93636	10·23391	30	9·76609	21	9·70245	44
17	8	9·93628	10·23361	29	9·76639	22	9·70267	43
18	7	9·93621	10·23332	29	9·76668	21	9·70288	42
19	7	9·93614	10·23303	28	9·76697	22	9·70310	41
20	8	9·93606	10·23275	29	9·76725	22	9·70332	40
21	7	9·93599	10·23246	29	9·76754	21	9·70353	39
22	7	9·93591	10·23217	29	9·76783	22	9·70375	38
23	8	9·93584	10·23188	29	9·76812	21	9·70396	37
24	7	9·93577	10·23159	29	9·76841	22	9·70418	36
25	7	9·93569	10·23130	29	9·76870	21	9·70439	35
26	8	9·93562	10·23101	29	9·76899	22	9·70461	34
27	7	9·93554	10·23072	29	9·76928	21	9·70482	33
28	7	9·93547	10·23043	29	9·76957	22	9·70504	32
29	8	9·93539	10·23014	29	9·76986	21	9·70525	31
30	7	9·93532	10·22985		9·77015	22	9·70547	30
′		Sine.	Tangent.		Cotang.		Cosine.	′

[59 degrees.]

[31 degrees.]   [58 degrees.]

′	D.	Cosine.	Cotang.	Diff.	Tangent.	Diff.	Sine.	′
30	8	9·93077	10·21268	28	9·78732	20	9·71809	30
29	8	9·93069	10·21240	29	9·78760	21	9·71829	31
28	8	9·93061	10·21211	28	9·78789	20	9·71850	32
27	7	9·93053	10·21183	28	9·78817	21	9·71870	33
26	8	9·93046	10·21155	29	9·78845	20	9·71891	34
25	8	9·93038	10·21126	28	9·78874	21	9·71911	35
24	8	9·93030	10·21098	28	9·78902	20	9·71932	36
23	8	9·93022	10·21070	29	9·78930	21	9·71952	37
22	8	9·93014	10·21041	28	9·78959	21	9·71973	38
21	7	9·93007	10·21013	28	9·78987	20	9·71994	39
20	8	9·92999	10·20985	28	9·79015	20	9·72014	40
19	8	9·92991	10·20957	29	9·79043	20	9·72034	41
18	8	9·92983	10·20928	28	9·79072	21	9·72055	42
17	7	9·92976	10·20900	28	9·79100	20	9·72075	43
16	8	9·92968	10·20872	28	9·79128	21	9·72096	44
15	8	9·92960	10·20844	29	9·79156	21	9·72116	45
14	8	9·92952	10·20815	28	9·79185	20	9·72137	46
13	8	9·92944	10·20787	28	9·79213	20	9·72157	47
12	7	9·92936	10·20759	28	9·79241	20	9·72177	48
11	8	9·92929	10·20731	28	9·79269	21	9·72198	49
10	8	9·92921	10·20703	29	9·79297	20	9·72218	50
9	8	9·92913	10·20674	28	9·79326	20	9·72238	51
8	8	9·92905	10·20646	2·8	9·79354	21	9·72259	52
7	7	9·92897	10·20618	28	9·79382	20	9·72279	53
6	8	9·92889	10·20590	28	9·79410	20	9·72299	54
5	8	9·92881	10·20562	28	9·79438	21	9·72320	55
4	8	9·92874	10·20534	29	9·79466	20	9·72340	56
3	8	9·92866	10·20505	28	9·79495	20	9·72360	57
2	8	9·92858	10·20477	28	9·79523	21	9·72381	58
1	8	9·92850	10·20449	28	9·79551	20	9·72401	59
0	8	9·92842	10·20421		9·79579		9·72421	60
′		Sine.	Tangent.		Cotang.		Cosine.	′

132

[31 degrees.]   [58 degrees.]

′	Sine.	Diff.	Tangent.	Diff.	Cotang.	D.	Cosine.	′
0	9·71184	21	9·77877	29	10·22123	8	9·93307	60
1	9·71205	21	9·77906	29	10·22094	8	9·93299	59
2	9·71226	21	9·77935	28	10·22065	7	9·93291	58
3	9·71247	21	9·77963	29	10·22037	8	9·93284	57
4	9·71268	21	9·77992	28	10·22008	7	9·93276	56
5	9·71289	21	9·78020	29	10·21980	8	9·93269	55
6	9·71310	21	9·78049	28	10·21951	8	9·93261	54
7	9·71331	21	9·78077	29	10·21923	7	9·93253	53
8	9·71352	21	9·78106	29	10·21894	8	9·93246	52
9	9·71373	20	9·78135	28	10·21865	8	9·93238	51
10	9·71393	21	9·78163	29	10·21837	7	9·93230	50
11	9·71414	21	9·78192	28	10·21808	8	9·93223	49
12	9·71435	21	9·78220	29	10·21780	8	9·93215	48
13	9·71456	21	9·78249	28	10·21751	7	9·93207	47
14	9·71477	21	9·78277	29	10·21723	8	9·93200	46
15	9·71498	21	9·78306	28	10·21694	8	9·93192	45
16	9·71519	20	9·78334	29	10·21666	7	9·93184	44
17	9·71539	21	9·78363	28	10·21637	8	9·93177	43
18	9·71560	21	9·78391	28	10·21609	8	9·93169	42
19	9·71581	21	9·78419	29	10·21581	7	9·93161	41
20	9·71602	20	9·78448	28	10·21552	8	9·93154	40
21	9·71622	21	9·78476	29	10·21524	8	9·93146	39
22	9·71643	21	9·78505	28	10·21495	7	9·93138	38
23	9·71664	21	9·78533	29	10·21467	8	9·93131	37
24	9·71685	20	9·78562	28	10·21438	8	9·93123	36
25	9·71705	21	9·78590	28	10·21410	7	9·93115	35
26	9·71726	21	9·78618	29	10·21382	8	9·93108	34
27	9·71747	20	9·78647	28	10·21353	8	9·93100	33
28	9·71767	21	9·78675	29	10·21325	8	9·93092	32
29	9·71788	21	9·78704	28	10·21296	7	9·93084	31
30	9·71809		9·78732		10·21268		9·93077	30
′	Cosine.		Cotang.		Tangent.		Sine.	′

[31 degrees.]   [58 degrees.]

[32 degrees.]   [57 degrees.]

'	D.	Cosine.	Cotang.	Diff.	Tangent.	Diff.	Sine.	'
30	8	9·92603	10·19581	28	9·80419	19	9·73022	30
29	8	9·92595	10·19553	28	9·80447	20	9·73041	31
28	8	9·92587	10·19526	27	9·80474	20	9·73061	32
27	8	9·92579	10·19498	28	9·80502	20	9·73081	33
26	8	9·92571	10·19470	28	9·80530	20	9·73101	34
25	8	9·92563	10·19442	28	9·80558	19	9·73121	35
24	9	9·92555	10·19414	28	9·80586	20	9·73140	36
23	8	9·92546	10·19386	28	9·80614	20	9·73160	37
22	8	9·92538	10·19358	28	9·80642	20	9·73180	38
21	8	9·92530	10·19331	27	9·80669	19	9·73200	39
20	8	9·92522	10·19303	28	9·80697	20	9·73219	40
19	8	9·92514	10·19275	28	9·80725	20	9·73239	41
18	8	9·92506	10·19247	28	9·80753	19	9·73259	42
17	8	9·92498	10·19219	28	9·80781	20	9·73278	43
16	8	9·92490	10·19192	27	9·80808	20	9·73298	44
15	9	9·92482	10·19164	28	9·80836	19	9·73318	45
14	8	9·92473	10·19136	28	9·80864	20	9·73337	46
13	8	9·92465	10·19108	28	9·80892	20	9·73357	47
12	8	9·92457	10·19081	27	9·80919	19	9·73377	48
11	8	9·92449	10·19053	28	9·80947	20	9·73396	49
10	8	9·92441	10·19025	28	9·80975	20	9·73416	50
9	8	9·92433	10·18997	28	9·81003	19	9·73435	51
8	9	9·92425	10·18970	27	9·81030	20	9·73455	52
7	8	9·92416	10·18942	28	9·81058	19	9·73474	53
6	8	9·92408	10·18914	28	9·81086	20	9·73494	54
5	8	9·92400	10·18887	27	9·81113	19	9·73513	55
4	8	9·92392	10·18859	28	9·81141	20	9·73533	56
3	8	9·92384	10·18831	28	9·81169	19	9·73552	57
2	9	9·92376	10·18804	27	9·81196	20	9·73572	58
1	8	9·92367	10·18776	28	9·81224	19	9·73591	59
0	8	9·92359	10·18748	28	9·81252	20	9·73611	60
'		Sine.	Tangent.		Cotang.		Cosine.	'

[32 degrees.]   [57 degrees.]

133

[32 degrees.]   [57 degrees.]

'	D.	Cosine.	Cotang.	Diff	Tangent.	Diff.	Sine.	'
0	8	9·92842	10·20421	28	9·79579	20	9·72421	60
1	8	9·92834	10·20393	28	9·79607	20	9·72441	59
2	8	9·92826	10·20365	28	9·79635	21	9·72461	58
3	8	9·92818	10·20337	28	9·79663	20	9·72482	57
4	8	9·92810	10·20309	28	9·79691	20	9·72502	56
5	7	9·92803	10·20281	28	9·79719	20	9·72522	55
6	8	9·92795	10·20253	29	9·79747	20	9·72542	54
7	8	9·92787	10·20224	28	9·79776	20	9·72562	53
8	8	9·92779	10·20196	28	9·79804	20	9·72582	52
9	8	9·92771	10·20168	28	9·79832	20	9·72602	51
10	8	9·92763	10·20140	28	9·79860	21	9·72622	50
11	8	9·92755	10·20112	28	9·79888	20	9·72643	49
12	8	9·92747	10·20084	28	9·79916	20	9·72663	48
13	8	9·92739	10·20056	28	9·79944	20	9·72683	47
14	8	9·92731	10·20028	28	9·79972	20	9·72703	46
15	8	9·92723	10·20000	28	9·80000	20	9·72723	45
16	8	9·92715	10·19972	28	9·80028	20	9·72743	44
17	8	9·92707	10·19944	28	9·80056	20	9·72763	43
18	8	9·92699	10·19916	28	9·80084	20	9·72783	42
19	8	9·92691	10·19888	28	9·80112	20	9·72803	41
20	8	9·92683	10·19860	28	9·80140	20	9·72823	40
21	8	9·92675	10·19832	27	9·80168	20	9·72843	39
22	8	9·92667	10·19805	28	9·80195	20	9·72863	38
23	8	9·92659	10·19777	28	9·80223	20	9·72883	37
24	8	9·92651	10·19749	28	9·80251	19	9·72902	36
25	8	9·92643	10·19721	28	9·80279	20	9·72922	35
26	8	9·92635	10·19693	28	9·80307	20	9·72942	34
27	8	9·92627	10·19665	28	9·80335	20	9·72962	33
28	8	9·92619	10·19637	28	9·80363	20	9·72982	32
29	8	9·92611	10·19609	28	9·80391	20	9·73002	31
30	8	9·92603	10·19581		9·80419		9·73022	30
'		Sine.	Tangent.		Cotang.		Cosine.	'

[33 degrees.]   **[56 degrees.]**

'	D.	Cosine.	Cotang.	Diff.	Tangent.	Diff.	Sine.	'
30	8	9.92111	10.17922	28	9.82078	19	9.74189	30
29	8	9.92102	10.17894	27	9.82106	19	9.74208	31
28	9	9.92094	10.17867	28	9.82133	19	9.74227	32
27	8	9.92086	10.17839	27	9.82161	19	9.74246	33
26	9	9.92077	10.17812	27	9.82188	19	9.74265	34
25	8	9.92069	10.17785	28	9.82215	19	9.74284	35
24	9	9.92060	10.17757	27	9.82243	19	9.74303	36
23	8	9.92052	10.17730	28	9.82270	19	9.74322	37
22	9	9.92044	10.17702	27	9.82298	19	9.74341	38
21	8	9.92035	10.17675	27	9.82325	19	9.74360	39
20	9	9.92027	10.17648	28	9.82352	19	9.74379	40
19	8	9.92018	10.17620	27	9.82380	19	9.74398	41
18	9	9.92010	10.17593	28	9.82407	19	9.74417	42
17	8	9.92002	10.17565	27	9.82435	19	9.74436	43
16	8	9.91993	10.17538	27	9.82462	19	9.74455	44
15	9	9.91985	10.17511	28	9.82489	19	9.74474	45
14	8	9.91976	10.17483	27	9.82517	19	9.74493	46
13	9	9.91968	10.17456	27	9.82544	19	9.74512	47
12	8	9.91959	10.17429	28	9.82571	18	9.74531	48
11	9	9.91951	10.17401	27	9.82599	19	9.74549	49
10	8	9.91942	10.17374	27	9.82626	19	9.74568	50
9	9	9.91934	10.17347	28	9.82653	19	9.74587	51
8	8	9.91925	10.17319	27	9.82681	19	9.74606	52
7	9	9.91917	10.17292	27	9.82708	19	9.74625	53
6	8	9.91908	10.17265	27	9.82735	18	9.74644	54
5	9	9.91900	10.17238	28	9.82762	19	9.74662	55
4	8	9.91891	10.17210	27	9.82790	19	9.74681	56
3	9	9.91883	10.17183	27	9.82817	19	9.74700	57
2	8	9.91874	10.17156	27	9.82844	18	9.74719	58
1	9	9.91866	10.17129	27	9.82871	19	9.74737	59
0	9	9.91857	10.17101	28	9.82899	19	9.74756	60
'		Sine.	Tangent.		Cotang.		Cosine.	'

[33 degrees.]   **[56 degrees.]**

'	Sine.	Diff.	Tangent.	Diff.	Cotang.	Cosine.	D.	'
0	9.73611	19	9.81252	27	10.18748	9.92359	8	60
1	9.73630	20	9.81279	28	10.18721	9.92351	8	59
2	9.73650	19	9.81307	28	10.18693	9.92343	8	58
3	9.73669	20	9.81335	27	10.18665	9.92335	9	57
4	9.73689	19	9.81362	28	10.18638	9.92326	8	56
5	9.73708	19	9.81390	28	10.18610	9.92318	8	55
6	9.73727	20	9.81418	27	10.18582	9.92310	8	54
7	9.73747	19	9.81445	28	10.18555	9.92302	9	53
8	9.73766	19	9.81473	27	10.18527	9.92293	8	52
9	9.73785	20	9.81500	28	10.18500	9.92285	8	51
10	9.73805	19	9.81528	28	10.18472	9.92277	8	50
11	9.73824	19	9.81556	27	10.18444	9.92269	9	49
12	9.73843	20	9.81583	28	10.18417	9.92260	8	48
13	9.73863	19	9.81611	27	10.18389	9.92252	8	47
14	9.73882	19	9.81638	28	10.18362	9.92244	9	46
15	9.73901	20	9.81666	27	10.18334	9.92235	8	45
16	9.73921	19	9.81693	28	10.18307	9.92227	8	44
17	9.73940	19	9.81721	27	10.18279	9.92219	8	43
18	9.73959	19	9.81748	28	10.18252	9.92211	9	42
19	9.73978	19	9.81776	27	10.18224	9.92202	8	41
20	9.73997	20	9.81803	28	10.18197	9.92194	8	40
21	9.74017	19	9.81831	27	10.18169	9.92186	9	39
22	9.74036	19	9.81858	28	10.18142	9.92177	8	38
23	9.74055	19	9.81886	27	10.18114	9.92169	8	37
24	9.74074	19	9.81913	28	10.18087	9.92161	9	36
25	9.74093	20	9.81941	27	10.18059	9.92152	8	35
26	9.74113	19	9.81968	28	10.18032	9.92144	8	34
27	9.74132	19	9.81996	27	10.18004	9.92136	9	33
28	9.74151	19	9.82023	28	10.17977	9.92127	8	32
29	9.74170	19	9.82051	27	10.17949	9.92119	8	31
30	9.74189		9.82078		10.17922	9.92111		30
'	Cosine.		Cotang.		Tangent.	Sine.		'

[34 degrees.]

'	Sine.	Diff.	Tangent.	Diff.	Cotang.	D.	Cosine.	'
30	9·75313	18	9·83713	27	10·16287	8	9·91599	30
31	9·75331	19	9·83740	28	10·16260	9	9·91591	29
32	9·75350	18	9·83768	27	10·16232	8	9·91582	28
33	9·75368	18	9·83795	27	10·16205	8	9·91573	27
34	9·75386	19	9·83822	27	10·16178	9	9·91565	26
35	9·75405	18	9·83849	27	10·16151	9	9·91556	25
36	9·75423	18	9·83876	27	10·16124	9	9·91547	24
37	9·75441	18	9·83903	27	10·16097	8	9·91538	23
38	9·75459	19	9·83930	27	10·16070	9	9·91530	22
39	9·75478	18	9·83957	27	10·16043	9	9·91521	21
40	9·75496	18	9·83984	27	10·16016	8	9·91512	20
41	9·75514	19	9·84011	27	10·15989	9	9·91504	19
42	9·75533	18	9·84038	27	10·15962	9	9·91495	18
43	9·75551	18	9·84065	27	10·15935	9	9·91486	17
44	9·75569	18	9·84092	27	10·15908	8	9·91477	16
45	9·75587	18	9·84119	27	10·15881	9	9·91469	15
46	9·75605	19	9·84146	27	10·15854	9	9·91460	14
47	9·75624	18	9·84173	27	10·15827	9	9·91451	13
48	9·75642	18	9·84200	27	10·15800	9	9·91442	12
49	9·75660	18	9·84227	27	10·15773	8	9·91433	11
50	9·75678	18	9·84254	26	10·15746	9	9·91425	10
51	9·75696	18	9·84280	27	10·15720	9	9·91416	9
52	9·75714	19	9·84307	27	10·15693	9	9·91407	8
53	9·75733	18	9·84334	27	10·15666	9	9·91398	7
54	9·75751	18	9·84361	27	10·15639	8	9·91389	6
55	9·75769	18	9·84388	27	10·15612	9	9·91381	5
56	9·75787	18	9·84415	27	10·15585	9	9·91372	4
57	9·75805	18	9·84442	27	10·15558	9	9·91363	3
58	9·75823	18	9·84469	27	10·15531	9	9·91354	2
59	9·75841	18	9·84496	27	10·15504	9	9·91345	1
60	9·75859		9·84523		10·15477	9	9·91336	0
'	Cosine.		Cotang.		Tangent.		Sine.	'

[55 degrees.]

135

[34 degrees.]

'	Sine.	Diff.	Tangent.	Diff.	Cotang.	D.	Cosine.	'
0	9·74756	19	9·82899	27	10·17101	8	9·91857	60
1	9·74775	19	9·82926	27	10·17074	8	9·91849	59
2	9·74794	18	9·82953	27	10·17047	8	9·91840	58
3	9·74812	19	9·82980	28	10·17020	9	9·91832	57
4	9·74831	19	9·83008	27	10·16992	8	9·91823	56
5	9·74850	18	9·83035	27	10·16965	9	9·91815	55
6	9·74868	19	9·83062	27	10·16938	8	9·91806	54
7	9·74887	19	9·83089	28	10·16911	9	9·91798	53
8	9·74906	18	9·83117	27	10·16883	8	9·91789	52
9	9·74924	19	9·83144	27	10·16856	9	9·91781	51
10	9·74943	18	9·83171	27	10·16829	9	9·91772	50
11	9·74961	19	9·83198	27	10·16802	8	9·91763	49
12	9·74980	19	9·83225	27	10·16775	9	9·91755	48
13	9·74999	18	9·83252	28	10·16748	8	9·91746	47
14	9·75017	19	9·83280	27	10·16720	9	9·91738	46
15	9·75036	18	9·83307	27	10·16693	9	9·91729	45
16	9·75054	19	9·83334	27	10·16666	8	9·91720	44
17	9·75073	18	9·83361	27	10·16639	9	9·91712	43
18	9·75091	19	9·83388	27	10·16612	8	9·91703	42
19	9·75110	18	9·83415	27	10·16585	9	9·91695	41
20	9·75128	19	9·83442	28	10·16558	9	9·91686	40
21	9·75147	18	9·83470	27	10·16530	8	9·91677	39
22	9·75165	19	9·83497	27	10·16503	9	9·91669	38
23	9·75184	18	9·83524	27	10·16476	9	9·91660	37
24	9·75202	19	9·83551	27	10·16449	8	9·91651	36
25	9·75221	18	9·83578	27	10·16422	9	9·91643	35
26	9·75239	19	9·83605	27	10·16395	9	9·91634	34
27	9·75258	18	9·83632	27	10·16368	8	9·91625	33
28	9·75276	18	9·83659	27	10·16341	9	9·91617	32
29	9·75294	19	9·83686	27	10·16314	9	9·91608	31
30	9·75313		9·83713		10·16287		9·91599	30
'	Cosine.		Cotang.		Tangent.		Sine.	'

[55 degrees.]

[35 degrees.]   **[54 degrees.]**

′	Diff.	Cosine.	Cotang.	Diff.	Tangent.	Diff.	Sine.	′
30	9	9·91069	10·14673	27	9·85327	18	9·76395	30
29	9	9·91060	10·14646	26	9·85354	18	9·76413	31
28	9	9·91051	10·14620	27	9·85380	17	9·76431	32
27	9	9·91042	10·14593	27	9·85407	18	9·76448	33
26	10	9·91033	10·14566	26	9·85434	18	9·76466	34
25	9	9·91023	10·14540	27	9·85460	17	9·76484	35
24	9	9·91014	10·14513	27	9·85487	18	9·76501	36
23	9	9·91005	10·14486	26	9·85514	18	9·76519	37
22	9	9·90996	10·14460	27	9·85540	17	9·76537	38
21	9	9·90987	10·14433	27	9·85567	18	9·76554	39
20	9	9·90978	10·14406	26	9·85594	18	9·76572	40
19	9	9·90969	10·14380	27	9·85620	17	9·76590	41
18	9	9·90960	10·14353	27	9·85647	18	9·76607	42
17	9	9·90951	10·14326	26	9·85674	17	9·76625	43
16	9	9·90942	10·14300	27	9·85700	18	9·76642	44
15	9	9·90933	10·14273	27	9·85727	18	9·76660	45
14	9	9·90924	10·14246	26	9·85754	17	9·76677	46
13	9	9·90915	10·14220	27	9·85780	18	9·76695	47
12	10	9·90906	10·14193	27	9·85807	17	9·76712	48
11	9	9·90896	10·14166	26	9·85834	18	9·76730	49
10	9	9·90887	10·14140	27	9·85860	17	9·76747	50
9	9	9·90878	10·14113	27	9·85887	18	9·76765	51
8	9	9·90869	10·14087	26	9·85913	17	9·76782	52
7	9	9·90860	10·14060	27	9·85940	18	9·76800	53
6	9	9·90851	10·14033	27	9·85967	17	9·76817	54
5	10	9·90842	10·14007	26	9·85993	18	9·76835	55
4	9	9·90832	10·13980	27	9·86020	17	9·76852	56
3	9	9·90823	10·13954	26	9·86046	18	9·76870	57
2	9	9·90814	10·13927	27	9·86073	17	9·76887	58
1	9	9·90805	10·13900	27	9·86100	18	9·76904	59
0		9·90796	10·13874	26	9·86126		9·76922	60
		Sine.	Tangent.		Cotang.		Cosine.	

[35 degrees.]   **[54 degrees.]**

136

[35 degrees.]   **[54 degrees.]**

′	Sine.	Diff.	Tangent.	Diff.	Cotang.	Cosine.	D.	′
0	9·75859	18	9·84523	27	10·15477	9·91336	8	60
1	9·75877	18	9·84550	26	10·15450	9·91328	9	59
2	9·75895	18	9·84576	27	10·15424	9·91319	9	58
3	9·75913	18	9·84603	27	10·15397	9·91310	9	57
4	9·75931	18	9·84630	27	10·15370	9·91301	9	56
5	9·75949	18	9·84657	27	10·15343	9·91292	9	55
6	9·75967	18	9·84684	27	10·15316	9·91283	9	54
7	9·75985	18	9·84711	27	10·15289	9·91274	8	53
8	9·76003	18	9·84738	26	10·15262	9·91266	9	52
9	9·76021	18	9·84764	27	10·15236	9·91257	9	51
10	9·76039	18	9·84791	27	10·15209	9·91248	9	50
11	9·76057	18	9·84818	27	10·15182	9·91239	9	49
12	9·76075	18	9·84845	27	10·15155	9·91230	9	48
13	9·76093	18	9·84872	27	10·15128	9·91221	9	47
14	9·76111	18	9·84899	26	10·15101	9·91212	9	46
15	9·76129	17	9·84925	27	10·15075	9·91203	9	45
16	9·76146	18	9·84952	27	10·15048	9·91194	9	44
17	9·76164	18	9·84979	27	10·15021	9·91185	9	43
18	9·76182	18	9·85006	27	10·14994	9·91176	9	42
19	9·76200	18	9·85033	26	10·14967	9·91167	9	41
20	9·76218	18	9·85059	27	10·14941	9·91158	9	40
21	9·76236	17	9·85086	27	10·14914	9·91149	8	39
22	9·76253	18	9·85113	27	10·14887	9·91141	9	38
23	9·76271	18	9·85140	26	10·14860	9·91132	9	37
24	9·76289	18	9·85166	27	10·14834	9·91123	9	36
25	9·76307	17	9·85193	27	10·14807	9·91114	9	35
26	9·76324	18	9·85220	27	10·14780	9·91105	9	34
27	9·76342	18	9·85247	26	10·14753	9·91096	9	33
28	9·76360	18	9·85273	27	10·14727	9·91087	9	32
29	9·76378	17	9·85300	27	10·14700	9·91078	9	31
30	9·76395		9·85327		10·14673	9·91069		30
	Cosine.		Cotang.		Tangent.	Sine.		

[35 degrees.]   **[54 degrees.]**

[36 degrees.] — [53 degrees.]

′	Diff.	Cosine.	Cotang.	Diff.	Tangent.	Diff.	Sine.	′
30	9	9·90518	10·13079	26	9·86921	17	9·77439	30
29	10	9·90509	10·13053	27	9·86947	17	9·77456	31
28	9	9·90499	10·13026	26	9·86974	17	9·77473	32
27	9	9·90490	10·13000	27	9·87000	17	9·77490	33
26	10	9·90480	10·12973	26	9·87027	17	9·77507	34
25	9	9·90471	10·12947	26	9·87053	17	9·77524	35
24	9	9·90462	10·12921	27	9·87079	17	9·77541	36
23	10	9·90452	10·12894	26	9·87106	17	9·77558	37
22	9	9·90443	10·12868	26	9·87132	17	9·77575	38
21	10	9·90434	10·12842	27	9·87158	17	9·77592	39
20	9	9·90424	10·12815	26	9·87185	17	9·77609	40
19	10	9·90415	10·12789	27	9·87211	17	9·77626	41
18	9	9·90405	10·12762	26	9·87238	17	9·77643	42
17	10	9·90396	10·12736	26	9·87264	17	9·77660	43
16	9	9·90386	10·12710	27	9·87290	17	9·77677	44
15	9	9·90377	10·12683	26	9·87317	17	9·77694	45
14	10	9·90368	10·12657	26	9·87343	17	9·77711	46
13	9	9·90358	10·12631	27	9·87369	16	9·77728	47
12	10	9·90349	10·12604	26	9·87396	17	9·77744	48
11	9	9·90339	10·12578	26	9·87422	17	9·77761	49
10	10	9·90330	10·12552	27	9·87448	17	9·77778	50
9	9	9·90320	10·12525	26	9·87475	17	9·77795	51
8	10	9·90311	10·12499	26	9·87501	17	9·77812	52
7	9	9·90301	10·12473	27	9·87527	17	9·77829	53
6	9	9·90292	10·12446	26	9·87554	16	9·77846	54
5	10	9·90282	10·12420	26	9·87580	17	9·77862	55
4	9	9·90273	10·12394	27	9·87606	17	9·77879	56
3	10	9·90263	10·12367	26	9·87633	17	9·77896	57
2	9	9·90254	10·12341	26	9·87659	17	9·77913	58
1	9	9·90244	10·12315	26	9·87685	16	9·77930	59
0	9	9·90235	10·12289		9·87711		9·77946	60
′		Sine.	Tangent.		Cotang.		Cosine.	′

[53 degrees.]

137

[36 degrees.] — [53 degrees.]

′	Sine.	Diff.	Tangent.	Diff.	Cotang.	Cosine.	Diff.	′
0	9·75922	17	9·86126	27	10·13874	9·90796	9	60
1	9·75939	18	9·86153	26	10·13847	9·90787	10	59
2	9·75957	17	9·86179	27	10·13821	9·90777	9	58
3	9·75974	17	9·86206	26	10·13794	9·90768	9	57
4	9·75991	18	9·86232	27	10·13768	9·90759	9	56
5	9·76009	17	9·86259	26	10·13741	9·90750	9	55
6	9·76026	17	9·86285	27	10·13715	9·90741	10	54
7	9·76043	18	9·86312	26	10·13688	9·90731	9	53
8	9·76061	17	9·86338	27	10·13662	9·90722	9	52
9	9·76078	17	9·86365	27	10·13635	9·90713	9	51
10	9·76095	17	9·86392	26	10·13608	9·90704	10	50
11	9·76112	18	9·86418	27	10·13582	9·90694	9	49
12	9·76130	17	9·86445	26	10·13555	9·90685	9	48
13	9·76147	17	9·86471	27	10·13529	9·90676	9	47
14	9·76164	17	9·86498	26	10·13502	9·90667	10	46
15	9·76181	18	9·86524	27	10·13476	9·90657	9	45
16	9·76199	17	9·86551	26	10·13449	9·90648	9	44
17	9·76216	17	9·86577	26	10·13423	9·90639	10	43
18	9·76233	17	9·86603	27	10·13397	9·90630	10	42
19	9·76250	18	9·86630	26	10·13370	9·90620	9	41
20	9·76268	17	9·86656	27	10·13344	9·90611	9	40
21	9·76285	17	9·86683	26	10·13317	9·90602	10	39
22	9·76302	17	9·86709	27	10·13291	9·90592	9	38
23	9·76319	17	9·86736	26	10·13264	9·90583	9	37
24	9·76336	17	9·86762	27	10·13238	9·90574	10	36
25	9·76353	17	9·86789	26	10·13211	9·90565	10	35
26	9·76370	17	9·86815	27	10·13185	9·90555	9	34
27	9·76387	18	9·86842	26	10·13158	9·90546	9	33
28	9·76405	17	9·86868	26	10·13132	9·90537	10	32
29	9·76422	17	9·86894	27	10·13106	9·90527	9	31
30	9·76439		9·86921		10·13079	9·90518	9	30
′	Cosine.		Cotang.		Tangent.	Sine.		′

[53 degrees.]

[37 degrees.]   **[52 degrees.]**

'	Sine.	Diff.	Tangent.	Diff.	Cotang.	Cosine.	Diff.	'
30	9·78445	16	9·88498	26	10·11502	9·89947	10	30
31	9·78461	17	9·88524	26	10·11476	9·89937	10	29
32	9·78478	16	9·88550	27	10·11450	9·89927	9	28
33	9·78494	16	9·88577	26	10·11423	9·89918	10	27
34	9·78510	17	9·88603	26	10·11397	9·89908	10	26
35	9·78527	16	9·88629	26	10·11371	9·89898	10	25
36	9·78543	17	9·88655	26	10·11345	9·89888	9	24
37	9·78560	16	9·88681	26	10·11319	9·89879	10	23
38	9·78576	16	9·88707	26	10·11293	9·89869	10	22
39	9·78592	17	9·88733	26	10·11267	9·89859	10	21
40	9·78609	16	9·88759	27	10·11241	9·89849	9	20
41	9·78625	17	9·88786	26	10·11214	9·89840	10	19
42	9·78642	16	9·88812	26	10·11188	9·89830	10	18
43	9·78658	16	9·88838	26	10·11162	9·89820	10	17
44	9·78674	17	9·88864	26	10·11136	9·89810	9	16
45	9·78691	16	9·88890	26	10·11110	9·89801	10	15
46	9·78707	16	9·88916	26	10·11084	9·89791	10	14
47	9·78723	16	9·88942	26	10·11058	9·89781	10	13
48	9·78739	17	9·88968	26	10·11032	9·89771	10	12
49	9·78756	16	9·88994	26	10·11006	9·89761	9	11
50	9·78772	16	9·89020	26	10·10980	9·89752	10	10
51	9·78788	17	9·89046	27	10·10954	9·89742	10	9
52	9·78805	16	9·89073	26	10·10927	9·89732	10	8
53	9·78821	16	9·89099	26	10·10901	9·89722	10	7
54	9·78837	16	9·89125	26	10·10875	9·89712	10	6
55	9·78853	16	9·89151	26	10·10849	9·89702	9	5
56	9·78869	17	9·89177	26	10·10823	9·89693	10	4
57	9·78886	16	9·89203	26	10·10797	9·89683	10	3
58	9·78902	16	9·89229	26	10·10771	9·89673	10	2
59	9·78918	16	9·89255	26	10·10745	9·89663	10	1
60	9·78934		9·89281		10·10719	9·89653		0
'	Cosine.		Cotang.		Tangent.	Sine.		'

[52 degrees.]

138

[37 degrees.]   **[52 degrees.]**

'	Sine.	Diff.	Tangent.	Diff.	Cotang.	Cosine.	Diff.	'
0	9·77946	17	9·87711	27	10·12289	9·90235	10	60
1	9·77963	17	9·87738	26	10·12262	9·90225	9	59
2	9·77980	17	9·87764	26	10·12236	9·90216	10	58
3	9·77997	16	9·87790	27	10·12210	9·90206	9	57
4	9·78013	17	9·87817	26	10·12183	9·90197	10	56
5	9·78030	17	9·87843	26	10·12157	9·90187	9	55
6	9·78047	16	9·87869	26	10·12131	9·90178	10	54
7	9·78063	17	9·87895	27	10·12105	9·90168	9	53
8	9·78080	17	9·87922	26	10·12078	9·90159	10	52
9	9·78097	16	9·87948	26	10·12052	9·90149	10	51
10	9·78113	17	9·87974	26	10·12026	9·90139	9	50
11	9·78130	17	9·88000	27	10·12000	9·90130	10	49
12	9·78147	16	9·88027	26	10·11973	9·90120	9	48
13	9·78163	17	9·88053	26	10·11947	9·90111	10	47
14	9·78180	17	9·88079	26	10·11921	9·90101	10	46
15	9·78197	16	9·88105	26	10·11895	9·90091	9	45
16	9·78213	17	9·88131	27	10·11869	9·90082	10	44
17	9·78230	16	9·88158	26	10·11842	9·90072	9	43
18	9·78246	17	9·88184	26	10·11816	9·90063	10	42
19	9·78263	17	9·88210	26	10·11790	9·90053	10	41
20	9·78280	16	9·88236	26	10·11764	9·90043	9	40
21	9·78296	17	9·88262	27	10·11738	9·90034	10	39
22	9·78313	16	9·88289	26	10·11711	9·90024	10	38
23	9·78329	17	9·88315	26	10·11685	9·90014	9	37
24	9·78346	16	9·88341	26	10·11659	9·90005	10	36
25	9·78362	17	9·88367	26	10·11633	9·89995	10	35
26	9·78379	16	9·88393	27	10·11607	9·89985	9	34
27	9·78395	17	9·88420	26	10·11580	9·89976	10	33
28	9·78412	16	9·88446	26	10·11554	9·89966	10	32
29	9·78428	17	9·88472	26	10·11528	9·89956	9	31
30	9·78445		9·88498		10·11502	9·89947		30
'	Cosine.		Cotang.		Tangent.	Sine.		'

[52 degrees.]

[38 degrees.]   **[51 degrees.]**

′	Diff.	Cosine.	Cotang.	Diff.	Tangent.	Diff.	Sine.	′
30	10	9.89354	10.09939	25	9.90061	16	9.79415	30
29	10	9.89344	10.09914	26	9.90086	16	9.79431	31
28	10	9.89334	10.09888	26	9.90112	16	9.79447	32
27	10	9.89324	10.09862	26	9.90138	16	9.79463	33
26	10	9.89314	10.09836	26	9.90164	15	9.79478	34
25	10	9.89304	10.09810	26	9.90190	16	9.79494	35
24	10	9.89294	10.09784	26	9.90216	16	9.79510	36
23	10	9.89284	10.09758	26	9.90242	16	9.79526	37
22	10	9.89274	10.09732	26	9.90268	16	9.79542	38
21	10	9.89264	10.09706	26	9.90294	16	9.79558	39
20	10	9.89254	10.09680	26	9.90320	15	9.79573	40
19	11	9.89244	10.09654	25	9.90346	16	9.79589	41
18	10	9.89233	10.09629	26	9.90371	16	9.79605	42
17	10	9.89223	10.09603	26	9.90397	16	9.79621	43
16	10	9.89213	10.09577	26	9.90423	15	9.79636	44
15	10	9.89203	10.09551	26	9.90449	16	9.79652	45
14	10	9.89193	10.09525	26	9.90475	16	9.79668	46
13	10	9.89183	10.09499	26	9.90501	15	9.79684	47
12	11	9.89173	10.09473	26	9.90527	16	9.79699	48
11	10	9.89162	10.09447	25	9.90553	16	9.79715	49
10	10	9.89152	10.09422	26	9.90578	15	9.79731	50
9	10	9.89142	10.09396	26	9.90604	16	9.79746	51
8	10	9.89132	10.09370	26	9.90630	16	9.79762	52
7	10	9.89122	10.09344	26	9.90656	15	9.79778	53
6	11	9.89112	10.09318	26	9.90682	16	9.79793	54
5	10	9.89101	10.09292	26	9.90708	16	9.79809	55
4	10	9.89091	10.09266	26	9.90734	15	9.79825	56
3	10	9.89081	10.09241	25	9.90759	16	9.79840	57
2	11	9.89071	10.09215	26	9.90785	16	9.79856	58
1	10	9.89060	10.09189	26	9.90811	16	9.79872	59
0		9.89050	10.09163		9.90837	15	9.79887	60
′		Sine.	Tangent.		Cotang.		Cosine.	′

[51 degrees.]

139

[38 degrees.]   **[51 degrees.]**

′	Diff.	Cosine.	Cotang.	Diff.	Tangent.	Diff.	Sine.	′
0	10	9.89653	10.10719	26	9.89281	16	9.78934	60
1	10	9.89643	10.10693	26	9.89307	17	9.78950	59
2	9	9.89633	10.10667	26	9.89333	16	9.78967	58
3	10	9.89624	10.10641	26	9.89359	16	9.78983	57
4	10	9.89614	10.10615	26	9.89385	16	9.78999	56
5	10	9.89604	10.10589	26	9.89411	16	9.79015	55
6	10	9.89594	10.10563	26	9.89437	16	9.79031	54
7	10	9.89584	10.10537	26	9.89463	16	9.79047	53
8	10	9.89574	10.10511	26	9.89489	16	9.79063	52
9	10	9.89564	10.10485	26	9.89515	16	9.79079	51
10	10	9.89554	10.10459	26	9.89541	16	9.79095	50
11	10	9.89544	10.10433	26	9.89567	17	9.79111	49
12	10	9.89534	10.10407	26	9.89593	16	9.79128	48
13	10	9.89524	10.10381	26	9.89619	16	9.79144	47
14	10	9.89514	10.10355	26	9.89645	16	9.79160	46
15	9	9.89504	10.10329	26	9.89671	16	9.79176	45
16	10	9.89495	10.10303	26	9.89697	16	9.79192	44
17	10	9.89485	10.10277	26	9.89723	16	9.79208	43
18	10	9.89475	10.10251	26	9.89749	16	9.79224	42
19	10	9.89465	10.10225	26	9.89775	16	9.79240	41
20	10	9.89455	10.10199	26	9.89801	16	9.79256	40
21	10	9.89445	10.10173	26	9.89827	16	9.79272	39
22	10	9.89435	10.10147	26	9.89853	16	9.79288	38
23	10	9.89425	10.10121	26	9.89879	15	9.79304	37
24	10	9.89415	10.10095	26	9.89905	16	9.79319	36
25	10	9.89405	10.10069	26	9.89931	16	9.79335	35
26	10	9.89395	10.10043	26	9.89957	16	9.79351	34
27	10	9.89385	10.10017	26	9.89983	16	9.79367	33
28	11	9.89375	10.09991	26	9.90009	16	9.79383	32
29	10	9.89364	10.09965	26	9.90035	16	9.79399	31
30		9.89354	10.09939		9.90061	16	9.79415	30
′		Sine.	Tangent.		Cotang.		Cosine.	′

[51 degrees.]

[39 degrees.]   [50 degrees.]

'	Diff.	Cosine.	Cotang.	Diff.	Tangent.	Diff.	Sine.	'
30	11	9·88741	10·08390	26	9·91610	15	9·80351	30
29	10	9·88730	10·08364	26	9·91636	16	9·80366	31
28	11	9·88720	10·08338	26	9·91662	15	9·80382	32
27	10	9·88709	10·08312	25	9·91688	15	9·80397	33
26	11	9·88699	10·08287	26	9·91713	15	9·80412	34
25	11	9·88688	10·08261	26	9·91739	16	9·80428	35
24	10	9·88678	10·08235	26	9·91765	15	9·80443	36
23	10	9·88668	10·08209	25	9·91791	15	9·80458	37
22	11	9·88657	10·08184	26	9·91816	15	9·80473	38
21	10	9·88647	10·08158	26	9·91842	16	9·80489	39
20	11	9·88636	10·08132	25	9·91868	15	9·80504	40
19	10	9·88626	10·08107	26	9·91893	15	9·80519	41
18	11	9·88615	10·08081	26	9·91919	15	9·80534	42
17	10	9·88605	10·08055	26	9·91945	16	9·80550	43
16	11	9·88594	10·08029	25	9·91971	15	9·80565	44
15	10	9·88584	10·08004	26	9·91996	15	9·80580	45
14	11	9·88573	10·07978	26	9·92022	15	9·80595	46
13	10	9·88563	10·07952	25	9·92048	15	9·80610	47
12	11	9·88552	10·07927	26	9·92073	16	9·80625	48
11	10	9·88542	10·07901	26	9·92099	15	9·80641	49
10	11	9·88531	10·07875	25	9·92125	15	9·80656	50
9	10	9·88521	10·07850	26	9·92150	15	9·80671	51
8	11	9·88510	10·07824	26	9·92176	15	9·80686	52
7	11	9·88499	10·07798	25	9·92202	16	9·80701	53
6	10	9·88489	10·07773	26	9·92227	15	9·80716	54
5	11	9·88478	10·07747	26	9·92253	15	9·80731	55
4	10	9·88468	10·07721	26	9·92279	15	9·80746	56
3	11	9·88457	10·07696	25	9·92304	16	9·80762	57
2	10	9·88447	10·07670	26	9·92330	15	9·80777	58
1	11	9·88436	10·07644	26	9·92356	15	9·80792	59
0		9·88425	10·07619	25	9·92381		9·80807	60
'		Sine.	Tangent.		Cotang.		Cosine.	'

[50 degrees.]

140

[39 degrees.]   [50 degrees.]

'	Diff.	Cosine.	Cotang.	Diff.	Tangent.	Diff.	Sine.	'
60	10	9·89050	10·09163	26	9·90837	16	9·79887	0
59	10	9·89040	10·09137	26	9·90863	15	9·79903	1
58	10	9·89030	10·09111	25	9·90889	15	9·79918	2
57	11	9·89020	10·09086	26	9·90914	16	9·79934	3
56	10	9·89009	10·09060	26	9·90940	15	9·79950	4
55	10	9·88999	10·09034	26	9·90966	15	9·79965	5
54	11	9·88989	10·09008	26	9·90992	16	9·79981	6
53	10	9·88978	10·08982	25	9·91018	15	9·79996	7
52	10	9·88968	10·08957	26	9·91043	16	9·80012	8
51	11	9·88958	10·08931	26	9·91069	15	9·80027	9
50	10	9·88948	10·08905	26	9·91095	16	9·80043	10
49	10	9·88937	10·08879	26	9·91121	15	9·80058	11
48	10	9·88927	10·08853	26	9·91147	16	9·80074	12
47	11	9·88917	10·08828	25	9·91172	15	9·80089	13
46	10	9·88906	10·08802	26	9·91198	16	9·80105	14
45	10	9·88896	10·08776	26	9·91224	15	9·80120	15
44	11	9·88886	10·08750	26	9·91250	16	9·80136	16
43	10	9·88875	10·08724	26	9·91276	15	9·80151	17
42	10	9·88865	10·08699	25	9·91301	15	9·80166	18
41	11	9·88855	10·08673	26	9·91327	16	9·80182	19
40	10	9·88844	10·08647	26	9·91353	15	9·80197	20
39	10	9·88834	10·08621	25	9·91379	16	9·80213	21
38	11	9·88824	10·08596	26	9·91404	15	9·80228	22
37	10	9·88813	10·08570	26	9·91430	16	9·80244	23
36	10	9·88803	10·08544	25	9·91456	15	9·80259	24
35	11	9·88793	10·08518	26	9·91482	15	9·80274	25
34	10	9·88782	10·08493	26	9·91507	16	9·80290	26
33	11	9·88772	10·08467	26	9·91533	15	9·80305	27
32	10	9·88761	10·08441	26	9·91559	15	9·80320	28
31	11	9·88751	10·08415	25	9·91585	16	9·80336	29
30	10	9·88741	10·08390		9·91610	15	9·80351	30
'		Cosine.	Tangent.		Cotang.		Sine.	'

[50 degrees.]

[40 degrees.]   [49 degrees.]

'	Diff.	Cosine.	Cotang.	Diff.	Tangent.	Diff.	Sine.	'
30	11	9·88105	10·06850	25	9·93150	15	9·81254	30
29	11	9·88094	10·06825	26	9·93175	15	9·81269	31
28	11	9·88083	10·06799	26	9·93201	15	9·81284	32
27	11	9·88072	10·06773	25	9·93227	15	9·81299	33
26	10	9·88061	10·06748	26	9·93252	14	9·81314	34
25	11	9·88051	10·06722	25	9·93278	15	9·81328	35
24	11	9·88040	10·06697	26	9·93303	15	9·81343	36
23	11	9·88029	10·06671	25	9·93329	14	9·81358	37
22	11	9·88018	10·06646	26	9·93354	15	9·81372	38
21	11	9·88007	10·06620	26	9·93380	15	9·81387	39
20	11	9·87996	10·06594	25	9·93406	15	9·81402	40
19	10	9·87985	10·06569	26	9·93431	14	9·81417	41
18	11	9·87975	10·06543	25	9·93457	15	9·81431	42
17	11	9·87964	10·06518	26	9·93482	15	9·81446	43
16	11	9·87953	10·06492	25	9·93508	14	9·81461	44
15	11	9·87942	10·06467	26	9·93533	15	9·81475	45
14	11	9·87931	10·06441	25	9·93559	15	9·81490	46
13	11	9·87920	10·06416	26	9·93584	14	9·81505	47
12	11	9·87909	10·06390	26	9·93610	15	9·81519	48
11	11	9·87898	10·06364	25	9·93636	15	9·81534	49
10	10	9·87887	10·06339	26	9·93661	14	9·81549	50
9	11	9·87877	10·06313	25	9·93687	15	9·81563	51
8	11	9·87866	10·06288	26	9·93712	15	9·81578	52
7	11	9·87855	10·06262	25	9·93738	14	9·81592	53
6	11	9·87844	10·06237	26	9·93763	15	9·81607	54
5	11	9·87833	10·06211	25	9·93789	15	9·81622	55
4	11	9·87822	10·06186	26	9·93814	14	9·81636	56
3	11	9·87811	10·06160	25	9·93840	15	9·81651	57
2	11	9·87800	10·06135	26	9·93865	14	9·81665	58
1	11	9·87789	10·06109	25	9·93891	15	9·81680	59
0		9·87778	10·06084		9·93916		9·81694	60
'		Sine.	Tangent.		Cotang.		Cosine.	'

[49 degrees.]

141

[40 degrees.]   [49 degrees.]

'	Diff.	Cosine.	Cotang.	Diff.	Tangent.	Diff.	Sine.	'
0	10	9·88425	10·07619	26	9·92381	15	9·80807	60
1	11	9·88415	10·07593	26	9·92407	15	9·80822	59
2	10	9·88404	10·07567	25	9·92433	15	9·80837	58
3	11	9·88394	10·07542	26	9·92458	15	9·80852	57
4	11	9·88383	10·07516	26	9·92484	15	9·80867	56
5	10	9·88373	10·07490	25	9·92510	15	9·80882	55
6	11	9·88362	10·07465	26	9·92535	15	9·80897	54
7	11	9·88351	10·07439	26	9·92561	15	9·80912	53
8	10	9·88340	10·07413	25	9·92587	15	9·80927	52
9	11	9·88330	10·07388	26	9·92612	15	9·80942	51
10	11	9·88319	10·07362	25	9·92638	15	9·80957	50
11	10	9·88308	10·07337	26	9·92663	15	9·80972	49
12	11	9·88298	10·07311	26	9·92689	15	9·80987	48
13	11	9·88287	10·07285	25	9·92715	15	9·81002	47
14	10	9·88276	10·07260	26	9·92740	15	9·81017	46
15	11	9·88266	10·07234	26	9·92766	15	9·81032	45
16	11	9·88255	10·07208	25	9·92792	14	9·81047	44
17	10	9·88244	10·07183	26	9·92817	15	9·81061	43
18	11	9·88234	10·07157	25	9·92843	15	9·81076	42
19	11	9·88223	10·07132	26	9·92868	15	9·81091	41
20	10	9·88212	10·07106	26	9·92894	15	9·81106	40
21	11	9·88201	10·07080	25	9·92920	15	9·81121	39
22	11	9·88191	10·07055	26	9·92945	15	9·81136	38
23	10	9·88180	10·07029	25	9·92971	15	9·81151	37
24	11	9·88169	10·07004	26	9·92996	15	9·81166	36
25	10	9·88158	10·06978	26	9·93022	14	9·81180	35
26	11	9·88148	10·06952	25	9·93048	15	9·81195	34
27	11	9·88137	10·06927	26	9·93073	15	9·81210	33
28	11	9·88126	10·06901	25	9·93099	15	9·81225	32
29	10	9·88115	10·06876	26	9·93124	14	9·81240	31
30		9·88105	10·06850		9·93150		9·81254	30
'		Sine.	Tangent.		Cotang.		Cosine.	'

[49 degrees.]

[41 degrees.] / **[48 degrees.]**

′	Diff.	Cosine.	Cotang.	Diff.	Tangent.	Diff.	Sine.	′
30	12	9·87446	10·05319	25	9·94681	15	9·82126	30
29	11	9·87434	10·05294	26	9·94706	14	9·82141	31
28	11	9·87423	10·05268	25	9·94732	14	9·82155	32
27	11	9·87412	10·05243	26	9·94757	14	9·82169	33
26	11	9·87401	10·05217	25	9·94783	15	9·82184	34
25	11	9·87390	10·05192	26	9·94808	14	9·82198	35
24	12	9·87378	10·05166	25	9·94834	14	9·82212	36
23	11	9·87367	10·05141	25	9·94859	14	9·82226	37
22	11	9·87356	10·05116	26	9·94884	14	9·82240	38
21	11	9·87345	10·05090	25	9·94910	15	9·82255	39
20	11	9·87334	10·05065	26	9·94935	14	9·82269	40
19	12	9·87322	10·05039	25	9·94961	14	9·82283	41
18	11	9·87311	10·05014	26	9·94986	14	9·82297	42
17	11	9·87300	10·04988	25	9·95012	14	9·82311	43
16	12	9·87288	10·04963	25	9·95037	15	9·82326	44
15	11	9·87277	10·04938	26	9·95062	14	9·82340	45
14	11	9·87266	10·04912	25	9·95088	14	9·82354	46
13	11	9·87255	10·04887	26	9·95113	14	9·82368	47
12	12	9·87243	10·04861	25	9·95139	14	9·82382	48
11	11	9·87232	10·04836	26	9·95164	14	9·82396	49
10	11	9·87221	10·04810	26	9·95190	14	9·82410	50
9	12	9·87209	10·04785	25	9·95215	14	9·82424	51
8	11	9·87198	10·04760	25	9·95240	15	9·82439	52
7	11	9·87187	10·04734	26	9·95266	14	9·82453	53
6	12	9·87175	10·04709	25	9·95291	14	9·82467	54
5	11	9·87164	10·04683	26	9·95317	14	9·82481	55
4	11	9·87153	10·04658	25	9·95342	14	9·82495	56
3	12	9·87141	10·04632	26	9·95368	14	9·82509	57
2	11	9·87130	10·04607	25	9·95393	14	9·82523	58
1	11	9·87119	10·04582	26	9·95418	14	9·82537	59
0	12	9·87107	10·04556	26	9·95444	14	9·82551	60
′		Sine.	Tangent.		Cotang.		Cosine.	′

[41 degrees.] / **[48 degrees.]**

′	Diff.	Cosine.	Cotang.	Diff.	Tangent.	Diff.	Sine.	′
0	11	9·87778	10·06084	26	9·93916	15	9·81694	60
1	11	9·87767	10·06058	26	9·93942	15	9·81709	59
2	11	9·87756	10·06033	25	9·93967	14	9·81723	58
3	11	9·87745	10·06007	26	9·93993	15	9·81738	57
4	11	9·87734	10·05982	25	9·94018	14	9·81752	56
5	11	9·87723	10·05956	26	9·94044	15	9·81767	55
6	11	9·87712	10·05931	25	9·94069	14	9·81781	54
7	11	9·87701	10·05905	26	9·94095	15	9·81796	53
8	11	9·87690	10·05880	25	9·94120	14	9·81810	52
9	11	9·87679	10·05854	26	9·94146	15	9·81825	51
10	11	9·87668	10·05829	25	9·94171	14	9·81839	50
11	11	9·87657	10·05803	26	9·94197	15	9·81854	49
12	11	9·87646	10·05778	25	9·94222	14	9·81868	48
13	11	9·87635	10·05752	26	9·94248	14	9·81882	47
14	11	9·87624	10·05727	25	9·94273	15	9·81897	46
15	11	9·87613	10·05701	26	9·94299	14	9·81911	45
16	12	9·87601	10·05676	25	9·94324	15	9·81926	44
17	11	9·87590	10·05650	26	9·94350	14	9·81940	43
18	11	9·87579	10·05625	25	9·94375	15	9·81955	42
19	11	9·87568	10·05599	26	9·94401	14	9·81969	41
20	11	9·87557	10·05574	25	9·94426	14	9·81983	40
21	11	9·87546	10·05548	26	9·94452	15	9·81998	39
22	11	9·87535	10·05523	25	9·94477	14	9·82012	38
23	11	9·87524	10·05497	26	9·94503	14	9·82026	37
24	12	9·87513	10·05472	25	9·94528	15	9·82041	36
25	12	9·87501	10·05446	26	9·94554	14	9·82055	35
26	11	9·87490	10·05421	25	9·94579	14	9·82069	34
27	11	9·87479	10·05396	26	9·94604	14	9·82084	33
28	11	9·87468	10·05370	25	9·94630	14	9·82098	32
29	11	9·87457	10·05345	26	9·94655	14	9·82112	31
30	11	9·87446	10·05319	26	9·94681	14	9·82126	30
′		Sine.	Tangent.		Cotang.		Cosine.	′

[42 degrees.]   **[47 degrees.]**

′	Diff.	Cosine.	Cotang.	Diff.	Tangent.	Diff.	Sine.	′
30	11	9·86763	10·03795	26	9·96205	14	9·82968	30
29	12	9·86752	10·03769	25	9·96231	14	9·82982	31
28	12	9·86740	10·03744	25	9·96256	14	9·82996	32
27	11	9·86728	10·03719	26	9·96281	14	9·83010	33
26	12	9·86717	10·03693	25	9·96307	13	9·83023	34
25	12	9·86705	10·03668	25	9·96332	14	9·83037	35
24	11	9·86694	10·03643	26	9·96357	14	9·83051	36
23	12	9·86682	10·03617	25	9·96383	14	9·83065	37
22	12	9·86670	10·03592	25	9·96408	13	9·83078	38
21	11	9·86659	10·03567	26	9·96433	14	9·83092	39
20	12	9·86647	10·03541	25	9·96459	14	9·83106	40
19	12	9·86635	10·03516	26	9·96484	14	9·83120	41
18	11	9·86624	10·03490	25	9·96510	13	9·83133	42
17	12	9·86612	10·03465	25	9·96535	14	9·83147	43
16	12	9·86600	10·03440	26	9·96560	14	9·83161	44
15	11	9·86589	10·03414	25	9·96586	13	9·83174	45
14	12	9·86577	10·03389	25	9·96611	14	9·83188	46
13	12	9·86565	10·03364	26	9·96636	14	9·83202	47
12	11	9·86554	10·03338	25	9·96662	13	9·83215	48
11	12	9·86542	10·03313	25	9·96687	14	9·83229	49
10	12	9·86530	10·03288	26	9·96712	13	9·83242	50
9	12	9·86518	10·03262	25	9·96738	14	9·83256	51
8	11	9·86507	10·03237	25	9·96763	14	9·83270	52
7	12	9·86495	10·03212	26	9·96788	13	9·83283	53
6	12	9·86483	10·03186	25	9·96814	14	9·83297	54
5	11	9·86472	10·03161	25	9·96839	13	9·83310	55
4	12	9·86460	10·03136	26	9·96864	14	9·83324	56
3	12	9·86448	10·03110	25	9·96890	14	9·83338	57
2	12	9·86436	10·03085	25	9·96915	13	9·83351	58
1	11	9·86425	10·03060	26	9·96940	14	9·83365	59
0	12	9·86413	10·03034	26	9·96966	13	9·83378	60
′		Sine.	Tangent.		Cotang.		Cosine.	′

[47 degrees.]

143

[42 degrees.]   **[47 degrees.]**

′	Diff.	Cosine.	Cotang.	Diff.	Tangent.	Diff.	Sine.	′
0	11	9·87107	10·04556	25	9·95444	14	9·82551	60
1	11	9·87096	10·04531	26	9·95469	14	9·82565	59
2	12	9·87085	10·04505	25	9·95495	14	9·82579	58
3	11	9·87073	10·04480	25	9·95520	14	9·82593	57
4	11	9·87062	10·04455	26	9·95545	14	9·82607	56
5	12	9·87050	10·04429	26	9·95571	14	9·82621	55
6	11	9·87039	10·04404	25	9·95596	14	9·82635	54
7	11	9·87028	10·04378	26	9·95622	14	9·82649	53
8	12	9·87016	10·04353	25	9·95647	14	9·82663	52
9	11	9·87005	10·04328	25	9·95672	14	9·82677	51
10	12	9·86993	10·04302	26	9·95698	14	9·82691	50
11	11	9·86982	10·04277	25	9·95723	14	9·82705	49
12	12	9·86970	10·04252	26	9·95748	14	9·82719	48
13	11	9·86959	10·04226	26	9·95774	14	9·82733	47
14	12	9·86947	10·04201	26	9·95799	14	9·82747	46
15	11	9·86936	10·04175	25	9·95825	14	9·82761	45
16	12	9·86924	10·04150	25	9·95850	14	9·82775	44
17	11	9·86913	10·04125	26	9·95875	13	9·82788	43
18	11	9·86902	10·04099	26	9·95901	14	9·82802	42
19	12	9·86890	10·04074	25	9·95926	14	9·82816	41
20	11	9·86879	10·04048	26	9·95952	14	9·82830	40
21	12	9·86867	10·04023	25	9·95977	14	9·82844	39
22	12	9·86855	10·03998	26	9·96002	14	9·82858	38
23	11	9·86844	10·03972	26	9·96028	13	9·82872	37
24	12	9·86832	10·03947	25	9·96053	14	9·82885	36
25	11	9·86821	10·03922	26	9·96078	14	9·82899	35
26	12	9·86809	10·03896	26	9·96104	14	9·82913	34
27	11	9·86798	10·03871	25	9·96129	14	9·82927	33
28	12	9·86786	10·03845	26	9·96155	14	9·82941	32
29	11	9·86775	10·03820	25	9·96180	13	9·82955	31
30	12	9·86763	10·03795	25	9·96205	13	9·82968	30
′		Sine.	Tangent.		Cotang.		Cosine.	′

[47 degrees.]

[43 degrees.]   [46 degrees.]

′	Diff.	Cosine.	Cotang.	Diff.	Tangent.	Diff.	Sine.	′
30	12	9·86056	10·02275	25	9·97725	14	9·83781	30
29	12	9·86044	10·02250	26	9·97750	13	9·83795	31
28	12	9·86032	10·02224	25	9·97776	13	9·83808	32
27	12	9·86020	10·02199	25	9·97801	13	9·83821	33
26	12	9·86008	10·02174	26	9·97826	14	9·83834	34
25	12	9·85996	10·02149	25	9·97851	13	9·83848	35
24	12	9·85984	10·02123	25	9·97877	13	9·83861	36
23	12	9·85972	10·02098	26	9·97902	13	9·83874	37
22	12	9·85960	10·02073	25	9·97927	14	9·83887	38
21	12	9·85948	10·02047	25	9·97953	13	9·83901	39
20	12	9·85936	10·02022	26	9·97978	13	9·83914	40
19	12	9·85924	10·01997	25	9·98003	13	9·83927	41
18	12	9·85912	10·01971	25	9·98029	13	9·83940	42
17	12	9·85900	10·01946	25	9·98054	14	9·83954	43
16	12	9·85888	10·01921	25	9·98079	13	9·83967	44
15	12	9·85876	10·01896	26	9·98104	13	9·83980	45
14	13	9·85864	10·01870	25	9·98130	13	9·83993	46
13	12	9·85851	10·01845	26	9·98155	14	9·84006	47
12	12	9·85839	10·01820	25	9·98180	13	9·84020	48
11	12	9·85827	10·01794	26	9·98206	13	9·84033	49
10	12	9·85815	10·01769	25	9·98231	13	9·84046	50
9	12	9·85803	10·01744	25	9·98256	13	9·84059	51
8	13	9·85791	10·01719	26	9·98281	13	9·84072	52
7	12	9·85779	10·01693	25	9·98307	13	9·84085	53
6	12	9·85766	10·01668	25	9·98332	13	9·84098	54
5	12	9·85754	10·01643	26	9·98357	13	9·84111	55
4	12	9·85742	10·01617	25	9·98383	14	9·84125	56
3	12	9·85730	10·01592	25	9·98408	13	9·84138	57
2	12	9·85718	10·01567	25	9·98433	13	9·84151	58
1	12	9·85706	10·01542	26	9·98458	13	9·84164	59
0	13	9·85693	10·01516	25	9·98484	13	9·84177	60
		Sine.	Tangent.		Cotang.		Cosine.	′

[46 degrees.]

144

[43 degrees.]   [46 degrees.]

′	Diff.	Cosine.	Cotang.	Diff.	Tangent.	Diff.	Sine.	′
60	12	9·86413	10·03034	25	9·96966	14	9·83378	0
59	12	9·86401	10·03009	25	9·96991	13	9·83392	1
58	12	9·86389	10·02984	26	9·97016	14	9·83405	2
57	11	9·86377	10·02958	25	9·97042	13	9·83419	3
56	12	9·86366	10·02933	25	9·97067	14	9·83432	4
55	12	9·86354	10·02908	26	9·97092	13	9·83446	5
54	12	9·86342	10·02882	25	9·97118	14	9·83459	6
53	12	9·86330	10·02857	25	9·97143	13	9·83473	7
52	12	9·86318	10·02832	25	9·97168	14	9·83486	8
51	11	9·86306	10·02807	26	9·97193	13	9·83500	9
50	12	9·86295	10·02781	25	9·97219	14	9·83513	10
49	12	9·86283	10·02756	25	9·97244	13	9·83527	11
48	12	9·86271	10·02731	26	9·97269	14	9·83540	12
47	12	9·86259	10·02705	25	9·97295	13	9·83554	13
46	12	9·86247	10·02680	25	9·97320	13	9·83567	14
45	12	9·86235	10·02655	26	9·97345	14	9·83581	15
44	12	9·86223	10·02629	25	9·97371	13	9·83594	16
43	11	9·86211	10·02604	25	9·97396	14	9·83608	17
42	12	9·86200	10·02579	26	9·97421	13	9·83621	18
41	12	9·86188	10·02553	25	9·97447	14	9·83634	19
40	12	9·86176	10·02528	25	9·97472	13	9·83648	20
39	12	9·86164	10·02503	26	9·97497	14	9·83661	21
38	12	9·86152	10·02477	25	9·97523	13	9·83674	22
37	12	9·86140	10·02452	25	9·97548	13	9·83688	23
36	12	9·86128	10·02427	25	9·97573	14	9·83701	24
35	12	9·86116	10·02402	26	9·97598	13	9·83715	25
34	12	9·86104	10·02376	25	9·97624	13	9·83728	26
33	12	9·86092	10·02351	25	9·97649	14	9·83741	27
32	12	9·86080	10·02326	26	9·97674	13	9·83755	28
31	12	9·86068	10·02300	25	9·97699	13	9·83768	29
30	12	9·86056	10·02275		9·97725		9·83781	30
		Sine.	Tangent.		Cotang.		Cosine.	′

[46 degrees.]

[44 degrees.]

′	Diff.	Cosine.	Cotang.	Diff.	Tangent.	Diff.	Sine.	′
30	12	9·85334	10·00758	25	9·99242	13	9·84566	30
29	13	9·85312	10·00733	25	9·99267	13	9·84579	31
28	12	9·85299	10·00707	26	9·99293	13	9·84592	32
27	12	9·85287	10·00682	25	9·99318	13	9·84605	33
26	13	9·85274	10·00657	25	9·99343	13	9·84618	34
25	12	9·85262	10·00632	25	9·99368	12	9·84630	35
24	12	9·85250	10·00606	26	9·99394	13	9·84643	36
23	13	9·85237	10·00581	25	9·99419	13	9·84656	37
22	12	9·85225	10·00556	25	9·99444	13	9·84669	38
21	13	9·85212	10·00531	26	9·99469	13	9·84682	39
20	12	9·85200	10·00505	25	9·99495	12	9·84694	40
19	13	9·85187	10·00480	25	9·99520	13	9·84707	41
18	12	9·85175	10·00455	25	9·99545	13	9·84720	42
17	13	9·85162	10·00430	26	9·99570	13	9·84733	43
16	12	9·85150	10·00404	25	9·99596	12	9·84745	44
15	13	9·85137	10·00379	25	9·99621	13	9·84758	45
14	12	9·85125	10·00354	26	9·99646	13	9·84771	46
13	13	9·85112	10·00328	25	9·99672	13	9·84784	47
12	12	9·85100	10·00303	25	9·99697	12	9·84796	48
11	13	9·85087	10·00278	25	9·99722	13	9·84809	49
10	13	9·85074	10·00253	26	9·99747	13	9·84822	50
9	12	9·85062	10·00227	25	9·99773	13	9·84835	51
8	13	9·85049	10·00202	25	9·99798	12	9·84847	52
7	12	9·85037	10·00177	26	9·99823	13	9·84860	53
6	13	9·85024	10·00152	26	9·99848	13	9·84873	54
5	12	9·85012	10·00126	25	9·99874	12	9·84885	55
4	13	9·84999	10·00101	25	9·99899	13	9·84898	56
3	13	9·84986	10·00076	25	9·99924	13	9·84911	57
2	12	9·84974	10·00051	26	9·99949	12	9·84923	58
1	13	9·84961	10·00025	25	9·99975	13	9·84936	59
0	12	9·84949	10·00000		10·00000		9·84949	60
′		Sine.	Tangent.		Cotang.		Cosine.	′

[45 degrees.]

145

[44 degrees.]

′	Diff.	Cosine.	Cotang.	Diff.	Tangent.	Diff.	Sine.	′
0	12	9·85693	10·01516	25	9·98484	13	9·84177	60
1	12	9·85681	10·01491	25	9·98509	13	9·84190	59
2	12	9·85669	10·01466	26	9·98534	13	9·84203	58
3	12	9·85657	10·01440	25	9·98560	13	9·84216	57
4	13	9·85645	10·01415	25	9·98585	13	9·84229	56
5	12	9·85632	10·01390	25	9·98610	13	9·84242	55
6	12	9·85620	10·01365	26	9·98635	14	9·84255	54
7	12	9·85608	10·01339	25	9·98661	13	9·84269	53
8	13	9·85596	10·01314	25	9·98686	13	9·84282	52
9	12	9·85583	10·01289	26	9·98711	13	9·84295	51
10	12	9·85571	10·01263	25	9·98737	13	9·84308	50
11	12	9·85559	10·01238	25	9·98762	13	9·84321	49
12	12	9·85547	10·01213	26	9·98787	13	9·84334	48
13	13	9·85534	10·01188	26	9·98812	13	9·84347	47
14	12	9·85522	10·01162	25	9·98838	13	9·84360	46
15	12	9·85510	10·01137	25	9·98863	13	9·84373	45
16	13	9·85497	10·01112	25	9·98888	12	9·84385	44
17	12	9·85485	10·01087	26	9·98913	13	9·84398	43
18	12	9·85473	10·01061	25	9·98939	13	9·84411	42
19	13	9·85460	10·01036	25	9·98964	13	9·84424	41
20	12	9·85448	10·01011	26	9·98989	13	9·84437	40
21	12	9·85436	10·00985	25	9·99015	13	9·84450	39
22	13	9·85423	10·00960	25	9·99040	13	9·84463	38
23	12	9·85411	10·00935	26	9·99065	13	9·84476	37
24	12	9·85399	10·00910	25	9·99090	13	9·84489	36
25	13	9·85386	10·00884	26	9·99116	13	9·84502	35
26	12	9·85374	10·00859	25	9·99141	13	9·84515	34
27	13	9·85361	10·00834	25	9·99166	13	9·84528	33
28	12	9·85349	10·00809	25	9·99191	12	9·84540	32
29	12	9·85337	10·00783	26	9·99217	13	9·84553	31
30	13	9·85324	10·00758	25	9·99242	13	9·84566	30
′		Sine.	Tangent.		Cotang.		Cosine.	′

[45 degrees.]

TABLES OF RIGHT ASCENSION, DECLINATION, AND ASCENSIONAL DIFFERENCE

	ARIES AND LIBRA		ASCENSIONAL DIFFERENCE		
			51N30	52N30	53N25
Deg.	Declin.	Rt. Ascen.	London	Birming'm	Liverpool
o	o '	o '	o '	o '	o '
0	0 0	0 0	0 0	0 0	0 0
1	0 24	0 55	0 30	0 31	0 32
2	0 48	1 50	1 0	1 2	1 4
3	1 12	2 45	1 30	1 33	1 37
4	1 36	3 40	2 0	2 4	2 9
5	1 59	4 35	2 30	2 35	2 41
6	2 23	5 30	3 0	3 6	3 13
7	2 47	6 26	3 30	3 37	3 45
8	3 10	7 21	4 0	4 8	4 17
9	3 34	8 16	4 30	4 39	4 49
10	3 58	9 11	5 0	5 10	5 21
11	4 21	10 7	5 30	5 41	5 58
12	4 45	11 2	6 0	6 12	6 25
13	5 8	11 58	6 30	6 43	6 57
14	5 31	12 53	7 0	7 14	7 29
15	5 55	13 49	7 29	7 45	8 1
16	6 18	14 44	7 59	8 16	8 33
17	6 41	15 40	8 29	8 46	9 5
18	7 4	16 36	8 58	9 17	9 37
19	7 27	17 32	9 28	9 48	10 8
20	7 49	18 28	9 57	10 18	10 40
21	8 12	19 24	10 27	10 49	11 12
22	8 34	20 20	10 56	11 19	11 43
23	8 57	21 17	11 26	11 49	12 15
24	9 19	22 18	11 55	12 20	12 46
25	9 41	23 10	12 24	12 50	13 17
26	10 3	24 6	12 58	13 20	13 49
27	10 24	25 3	13 22	13 50	14 20
28	10 46	26 0	13 51	14 20	14 51
29	11 7	26 57	14 20	14 50	15 22
30	11 29	27 55	14 48	15 19	15 53

☞ For the R.A. of Libra add 180° to the same degree of Aries.
The Declin. and Asc. Diff. are the same for both.

TAURUS AND SCORPIO			ASCENSIONAL DIFFERENCE		
			51N30	52N30	53N25
Deg.	Declin.	Rt. Ascen.	London	Birming'm	Liverpool
o	o '	o '	o '	o '	o '
0	11 29	27 55	14 48	15 19	15 53
1	11 50	28 52	15 17	15 49	16 23
2	12 10	29 49	15 45	16 19	16 54
3	12 31	30 47	16 14	16 48	17 24
4	12 51	31 45	16 42	17 17	17 55
5	13 12	32 43	17 10	17 46	18 25
6	13 32	33 41	17 38	18 15	18 55
7	13 51	34 39	18 5	18 44	19 25
8	14 11	35 38	18 33	19 12	19 54
9	14 30	36 37	19 0	19 41	20 24
10	14 49	37 35	19 27	20 9	20 53
11	15 8	38 34	19 54	20 37	21 22
12	15 27	39 33	20 21	21 5	21 51
13	15 45	40 33	20 47	21 32	22 20
14	16 3	41 32	21 14	21 59	22 48
15	16 21	42 32	21 40	22 26	23 16
16	16 38	43 32	22 5	22 53	23 44
17	16 55	44 32	22 81	23 20	24 12
18	17 12	45 32	22 56	23 46	24 39
19	17 29	46 33	23 21	24 12	25 6
20	17 45	47 33	23 46	24 37	25 33
21	18 1	48 34	24 10	25 3	25 59
22	18 17	49 35	24 34	25 28	26 25
23	18 32	50 36	24 57	25 52	26 51
24	18 47	51 37	25 21	26 16	27 16
25	19 1	52 39	25 48	26 40	27 41
26	19 16	53 40	26 6	27 4	28 5
27	19 30	54 42	26 28	27 27	28 29
28	19 43	55 44	26 49	27 49	28 58
29	19 57	56 47	27 11	28 11	29 16
30	20 10	57 49	27 31	28 33	29 39

☞ For R.A. of Scorpio add 180° to the same degree of Taurus.
The Declin. and Asc. Diff. are the same for both.

			ASCENSIONAL DIFFERENCE		
GEMINI AND SAGITTARIUS			51N30	52N30	53N25
Deg.	Declin.	Rt. Ascen.	London	Birming'm	Liverpool
o	o '	o '	o '	o '	o '
0	20 10	57 49	27 31	28 33	29 39
1	20 22	58 52	27 52	28 54	30 1
2	20 35	59 54	28 12	29 15	30 23
3	20 46	60 57	28 31	29 35	30 44
4	20 57	62 0	28 49	29 54	31 4
5	21 8	63 3	29 8	30 13	31 24
6	21 19	64 7	29 25	30 32	31 43
7	21 29	65 10	29 42	30 50	32 2
8	21 39	66 14	29 59	31 7	32 20
9	21 49	67 18	30 15	31 23	32 37
10	21 58	68 22	30 30	31 40	32 54
11	22 6	69 26	30 45	31 55	33 10
12	22 14	70 30	30 58	32 9	33 26
13	22 22	71 34	31 11	32 23	33 40
14	22 29	72 39	31 24	32 37	33 54
15	22 36	73 43	31 36	32 49	34 7
16	22 48	74 48	31 48	33 1	34 20
17	22 49	75 52	31 58	33 12	34 31
18	22 55	76 57	32 8	33 22	34 42
19	23 0	78 2	32 17	33 32	34 52
20	23 4	79 7	32 25	33 41	35 1
21	23 9	80 12	32 33	33 49	35 10
22	23 13	81 17	32 40	33 56	35 17
23	23 16	82 22	32 46	34 2	35 24
24	23 19	83 28	32 51	34 7	35 30
25	23 21	84 33	32 55	34 12	35 35
26	23 23	85 38	32 59	34 16	35 39
27	23 25	86 44	33 2	34 19	35 42
28	23 26	87 49	33 4	34 21	35 44
29	23 27	88 55	33 5	34 22	35 45
30	23 27	90 0	33 6	34 23	35 46

☞ For the R.A. of Sagittarius add 180° to the same degree of Gemini.
The Declin. and Asc. Diff. are the same for both.

CANCER AND CAPRICORNUS			ASCENSIONAL DIFFERENCE		
			51N30	52N30	53N25
Deg.	Declin.	Rt. Ascen.	London	Birming'm	Liverpool
o	o '	o '	o '	o '	o '
0	23 27	90 0	33 6	34 23	35 46
1	23 27	91 5	33 5	34 22	35 45
2	23 26	92 11	33 4	34 21	35 44
3	23 25	93 16	33 2	34 19	35 42
4	23 23	94 22	32 59	34 16	35 39
5	23 21	95 27	32 55	34 12	35 35
6	23 19	96 32	32 51	34 7	35 30
7	23 16	97 38	32 46	34 2	35 24
8	23 13	98 43	32 40	33 56	35 17
9	23 9	99 48	32 33	33 49	35 10
10	23 4	100 53	32 25	33 41	35 1
11	23 0	101 58	32 17	33 32	34 52
12	22 55	103 3	32 8	33 22	34 42
13	22 49	104 8	31 58	33 12	34 31
14	22 43	105 12	31 48	33 1	34 20
15	22 36	106 17	31 36	32 49	34 7
16	22 29	107 21	31 24	32 37	33 54
17	22 22	108 26	31 11	32 23	33 40
18	22 14	109 30	30 58	32 9	33 26
19	22 6	110 34	30 45	31 55	33 10
20	21 58	111 38	30 30	31 40	32 54
21	21 49	112 42	30 15	31 23	32 37
22	21 39	113 46	29 59	31 7	32 20
23	21 29	114 50	29 42	30 50	32 2
24	21 19	115 53	29 25	30 32	31 43
25	21 8	116 57	29 8	30 13	31 24
26	20 57	118 0	28 49	29 54	31 4
27	20 46	119 3	28 31	29 35	30 44
28	20 35	120 6	28 12	29 16	30 23
29	20 22	121 8	27 52	28 54	30 1
30	20 10	122 11	27 31	28 33	29 39

☞ For the R.A. of Capricornus add 180° to the same degree of Cancer.

The Declin. and Asc. Diff. are the same for both.

LEO AND AQUARIUS			ASCENSIONAL DIFFERENCE		
			51N30	52N30	53N25
Deg.	Declin.	Rt. Ascen.	London	Birming'm	Liverpool
o	o '	o '	o '	o '	o '
0	20 10	122 11	27 31	28 33	29 39
1	19 57	123 13	27 11	28 11	29 16
2	19 43	124 16	26 49	27 49	28 53
3	19 30	125 18	26 28	27 27	28 29
4	19 16	126 20	26 6	27 4	28 5
5	19 1	127 21	25 43	26 40	27 41
6	18 47	128 23	25 21	26 16	27 16
7	18 32	129 24	24 57	25 52	26 51
8	18 17	130 25	24 34	25 28	26 25
9	18 1	131 26	24 10	25 3	25 59
10	17 45	132 27	23 46	24 37	25 33
11	17 29	133 27	23 21	24 12	25 6
12	17 12	134 28	22 56	23 46	24 39
13	16 55	135 28	22 31	23 20	24 12
14	16 38	136 28	22 5	22 53	23 44
15	16 21	137 28	21 40	22 26	23 16
16	16 3	138 28	21 14	21 59	22 48
17	15 45	139 27	20 47	21 32	22 20
18	15 27	140 27	20 21	21 5	21 51
19	15 8	141 26	19 54	20 37	21 22
20	14 49	142 25	19 27	20 9	20 53
21	14 30	143 23	19 0	19 41	20 24
22	14 11	144 22	18 33	19 12	19 54
23	13 51	145 21	18 5	18 44	19 25
24	13 32	146 19	17 38	18 15	18 55
25	13 12	147 17	17 10	17 46	18 25
26	12 51	148 15	16 42	17 17	17 55
27	12 31	149 13	16 14	16 48	17 24
28	12 10	150 11	15 45	16 19	16 54
29	11 50	151 8	15 17	15 49	16 23
30	11 29	152 5	14 48	15 19	15 53

☞ For the R.A. of Aquarius add 180° to the same degree of Leo.
The Declin. and Asc. Diff. are the same for both.

VIRGO AND PISCES			ASCENSIONAL DIFFERENCE		
			51N30	52N30	53N25
Deg.	Declin.	Rt. Ascen.	London	Birming'm	Liverpool
o	o '	o '	o '	o '	o '
0	11 29	152 5	14 48	15 19	15 53
1	11 7	153 3	14 20	14 50	15 22
2	10 46	154 0	13 51	14 20	14 51
3	10 24	154 57	13 22	13 50	14 20
4	10 3	155 54	12 53	13 20	13 49
5	9 41	156 50	12 24	12 50	13 17
6	9 19	157 47	11 55	12 20	12 46
7	8 57	158 43	11 26	11 49	12 15
8	8 34	159 40	10 56	11 19	11 43
9	8 12	160 36	10 27	10 49	11 12
10	7 49	161 32	9 57	10 18	10 40
11	7 27	162 28	9 28	9 48	10 8
12	7 4	163 24	8 58	9 17	9 37
13	6 41	164 20	8 29	8 46	9 5
14	6 18	165 16	7 59	8 16	8 33
15	5 55	166 11	7 29	7 45	8 1
16	5 31	167 7	7 0	7 14	7 29
17	5 8	168 2	6 30	6 43	6 57
18	4 45	168 58	6 0	6 12	6 25
19	4 21	169 53	5 30	5 41	5 53
20	3 58	170 49	5 0	5 10	5 21
21	3 34	171 44	4 30	4 39	4 49
22	3 10	172 39	4 0	4 8	4 17
23	2 47	173 34	3 30	3 37	3 45
24	2 23	174 30	3 0	3 6	3 13
25	1 59	175 25	2 30	2 35	2 41
26	1 36	176 20	2 0	2 4	2 9
27	1 12	177 15	1 30	1 33	1 37
28	0 48	178 10	1 0	1 2	1 4
29	0 24	179 5	0 30	0 31	0 32
30	0 0	180 0	0 0	0 0	0 0

☞ For the R.A. of Pisces add 180° to the same degree of Virgo.
The Declin. and Asc. Diff. are the same for both.

TERNARY PROPORTIONAL LOGARITHMS

′	0°	1°	2°	3°	4°	5°	6°	7°	8°	9°
0	Infinite	2·25527	1·95424	1·77815	1·65321	1·55630	1·47712	1·41017	1·35218	1·30103
1	4·03342	2·24809	1·95064	1·77575	1·65141	1·55486	1·47592	1·40914	1·35128	1·30023
2	3·73239	2·24103	1·94706	1·77335	1·64961	1·55342	1·47472	1·40811	1·35038	1·29942
3	3·55630	2·23408	1·94352	1·77097	1·64782	1·55198	1·47352	1·40708	1·34948	1·29862
4	3·43136	2·22724	1·94000	1·76861	1·64603	1·55055	1·47232	1·40606	1·34858	1·29782
5	3·33445	2·22051	1·93651	1·76625	1·64426	1·54912	1·47113	1·40503	1·34768	1·29703
6	3·25527	2·21388	1·93305	1·76391	1·64249	1·54770	1·46994	1·40401·4	1·34679	1·29623
7	3·18833	2·20735	1·92962	1·76158	1·64073	1·54629	1·46876	1·40300	1·34589	1·29544
8	3·13033	2·20091	1·92621	1·75927	1·63897	1·54487	1·46758	1·40198	1·34500	1·29464
9	3·07918	2·19457	1·92283	1·75696	1·63722	1·54347	1·46640	1·40097	1·34411	1·29385
10	3·03342	2·18833	1·91948	1·75467	1·63548	1·54206	1·46522	1·39996	1·34323	1·29306
11	2·99203	2·18217	1·91615	1·75239	1·63375	1·54066	1·46404	1·39895	1·34234	1·29227
12	2·95424	2·17609	1·91285	1·75012	1·63202	1·53927	1·46288	1·39794	1·34146	1·29148
13	2·91948	2·17010	1·90957	1·74787	1·63030	1·53788	1·46171	1·39694	1·34058	1·29070
14	2·88730	2·16419	1·90632	1·74562	1·62859	1·53649	1·46055	1·39593	1·33970	1·28991
15	2·85733	2·15836	1·90309	1·74339	1·62688	1·53511	1·45938	1·39493	1·33882	1·28913
16	2·82930	2·15261	1·89988	1·74117	1·62518	1·53374	1·45824	1·39394	1·33794	1·28835
17	2·80297	2·14693	1·89670	1·73896	1·62349	1·53236	1·45708	1·39294	1·33707	1·28757
18	2·77815	2·14133	1·89354	1·73676	1·62180	1·53100	1·45593	1·39195	1·33619	1·28679
19	2·75467	2·13580	1·89041	1·73457	1·62012	1·52963	1·45478	1·39096	1·33532	1·28601
20	2·73239	2·13033	1·88730	1·73239	1·61845	1·52827	1·45364	1·38997	1·33445	1·28524
21	2·71120	2·12494	1·88420	1·73023	1·61678	1·52692	1·45250	1·38899	1·33359	1·28446
22	2·69100	2·11961	1·88114	1·72807	1·61512	1·52557	1·45136	1·38800	1·33272	1·28369
23	2·67170	2·11435	1·87809	1·72593	1·61347	1·52422	1·45022	1·38702	1·33186	1·28292
24	2·65321	2·10914	1·87506	1·72379	1·61182	1·52288	1·44909	1·38604	1·33099	1·28215
25	2·63548	2·10400	1·87206	1·72167	1·61018	1·52154	1·44796	1·38506	1·33013	1·28138
26	2·61845	2·09893	1·86907	1·71956	1·60854	1·52021	1·44684	1·38409	1·32927	1·28061
27	2·60206	2·09390	1·86611	1·71745	1·60691	1·51888	1·44571	1·38312	1·32842	1·27984
28	2·58627	2·08894	1·86316	1·71536	1·60529	1·51755	1·44459	1·38215	1·32756	1·27908
29	2·57103	2·08403	1·86024	1·71328	1·60367	1·51623	1·44347	1·38118	1·32671	1·27831
30	2·55630	2·07918	1·85733	1·71120	1·60206	1·51491	1·44236	1·38021	1·32585	1·27755
31	2·54206	2·07438	1·85445	1·70914	1·60045	1·51360	1·44125	1·37925	1·32500	1·27679
32	2·52827	2·06964	1·85158	1·70709	1·59885	1·51229	1·44014	1·37829	1·32415	1·27603
33	2·51491	2·06494	1·84873	1·70504	1·59726	1·51098	1·43903	1·37733	1·32331	1·27527
34	2·50194	2·06030	1·84590	1·70301	1·59567	1·50968	1·43793	1·37637	1·32246	1·27451
35	2·48936	2·05570	1·84309	1·70099	1·59409	1·50838	1·43683	1·37541	1·32162	1·27376
36	2·47712	2·05115	1·84030	1·69897	1·59251	1·50708	1·43573	1·37446	1·32077	1·27300
37	2·46522	2·04665	1·83752	1·69696	1·59094	1·50579	1·43463	1·37351	1·31993	1·27225
38	2·45364	2·04220	1·83477	1·69497	1·58938	1·50451	1·43354	1·37256	1·31909	1·27150
39	2·44236	2·03779	1·83203	1·69298	1·58782	1·50322	1·43245	1·37161	1·31826	1·27075
40	2·43136	2·03342	1·82930	1·69100	1·58627	1·50194	1·43136	1·37067	1·31742	1·27000
41	2·42064	2·02910	1·82660	1·68903	1·58472	1·50067	1·43028	1·36972	1·31659	1·26925
42	2·41017	2·02482	1·82391	1·68707	1·58317	1·49940	1·42920	1·36878	1·31575	1·26850
43	2·39996	2·02060	1·82124	1·68512	1·58164	1·49813	1·42812	1·36784	1·31492	1·26776
44	2·38997	2·01639	1·81858	1·68318	1·58011	1·49687	1·42704	1·36691	1·31409	1·26701
45	2·38021	2·01223	1·81594	1·68124	1·57858	1·49560	1·42597	1·36597	1·31326	1·26627
46	2·37067	2·00812	1·81332	1·67932	1·57706	1·49435	1·42490	1·36504	1·31244	1·26553
47	2·36133	2·00404	1·81071	1·67740	1·57554	1·49309	1·42383	1·36411	1·31161	1·26479
48	2·35218	2·00000	1·80811	1·67549	1·57403	1·49184	1·42276	1·36318	1·31079	1·26405
49	2·34323	1·99600	1·80554	1·67359	1·57253	1·49060	1·42170	1·36225	1·30997	1·26331
50	2·33445	1·99203	1·80297	1·67170	1·57103	1·48936	1·42064	1·36133	1·30915	1·26257
51	2·32585	1·98810	1·80043	1·66981	1·56953	1·48812	1·41958	1·36040	1·30833	1·26184
52	2·31742	1·98421	1·79790	1·66794	1·56804	1·48688	1·41853	1·35948	1·30751	1·26110
53	2·30915	1·98035	1·79538	1·66607	1·56656	1·48565	1·41747	1·35856	1·30670	1·26037
54	2·30103	1·97652	1·79287	1·66421	1·56508	1·48442	1·41642	1·35765	1·30588	1·25964
55	2·29306	1·97273	1·79039	1·66236	1·56360	1·48320	1·41538	1·35673	1·30507	1·25891
56	2·28524	1·96897	1·78791	1·66051	1·56213	1·48197	1·41433	1·35582	1·30426	1·25818
57	2·27755	1·96524	1·78545	1·65868	1·56067	1·48076	1·41329	1·35491	1·30345	1·25745
58	2·27000	1·96154	1·78300	1·65685	1·55921	1·47954	1·41225	1·35400	1·30264	1·25672
59	2·26257	1·95788	1·78057	1·65503	1·55775	1·47813	1·41121	1·35309	1·30183	1·25600
60	2·25527	1·95424	1·77815	1·65321	1·55630	1·47712	1·41017	1·35218	1·30103	1·25527

157

′	10°	11°	12°	13°	14°	15°	16°	17°	18°	19°
0	1·25527	1·21388	1·17609	1·14133	1·10914	1·07918	1·05115	1·02482	1·00000	0·97652
1	1·25455	1·21322	1·17549	1·14077	1·10863	1·07870	1·05070	1·02440	0·99960	0·97614
2	1·25383	1·21257	1·17489	1·14022	1·10811	1·07822	1·05025	1·02397	0·99920	0·97576
3	1·25311	1·21191	1·17429	1·13966	1·10760	1·07774	1·04980	1·02355	0·99880	0·97538
4	1·25239	1·21126	1·17369	1·13911	1·10708	1·07726	1·04935	1·02312	0·99839	0·97500
5	1·25167	1·21060	1·17309	1·13855	1·10657	1·07678	1·04890	1·02270	0·99799	0·97462
6	1·25095	1·20995	1·17249	1·13800	1·10605	1·07630	1·04845	1·02228	0·99759	0·97424
7	1·25024	1·20930	1·17189	1·13745	1·10554	1·07582	1·04800	1·02185	0·99719	0·97386
8	1·24952	1·20865	1·17129	1·13690	1·10503	1·07534	1·04755	1·02143	0·99679	0·97348
9	1·24881	1·20800	1·17070	1·13635	1·10452	1·07486	1·04710	1·02101	0·99640	0·97310
10	1·24809	1·20735	1·17010	1·13580	1·10400	1·07438	1·04665	1·02059	0·99600	0·97273
11	1·24738	1·20670	1·16951	1·13525	1·10349	1·07391	1·04620	1·02017	0·99560	0·97235
12	1·24667	1·20605	1·16891	1·13470	1·10298	1·07343	1·04576	1·01974	0·99520	0·97197
13	1·24596	1·20541	1·16832	1·13415	1·10247	1·07295	1·04531	1·01932	0·99480	0·97159
14	1·24526	1·20476	1·16773	1·13360	1·10197	1·07248	1·04486	1·01890	0·99441	0·97122
15	1·24455	1·20412	1·16714	1·13306	1·10146	1·07200	1·04442	1·01848	0·99401	0·97084
16	1·24384	1·20348	1·16655	1·13251	1·10095	1·07153	1·04397	1·01806	0·99361	0·97047
17	1·24314	1·20284	1·16596	1·13197	1·10044	1·07105	1·04353	1·01764	0·99322	0·97009
18	1·24244	1·20219	1·16537	1·13142	1·09994	1·07058	1·04308	1·01723	0·99282	0·96972
19	1·24173	1·20155	1·16478	1·13088	1·09943	1·07011	1·04264	1·01681	0·99243	0·96934
20	1·24103	1·20091	1·16419	1·13033	1·09893	1·06964	1·04220	1·01639	0·99203	0·96897
21	1·24033	1·20028	1·16361	1·12979	1·09842	1·06916	1·04175	1·01597	0·99164	0·96859
22	1·23963	1·19964	1·16302	1·12925	1·09792	1·06869	1·04131	1·01556	0·99124	0·96822
23	1·23894	1·19900	1·16243	1·12871	1·09741	1·06822	1·04087	1·01514	0·99085	0·96784
24	1·23824	1·19837	1·16185	1·12817	1·09691	1·06775	1·04043	1·01472	0·99045	0·96747
25	1·23754	1·19773	1·16127	1·12763	1·09641	1·06728	1·03999	1·01431	0·99006	0·96710
26	1·23685	1·19710	1·16068	1·12709	1·09591	1·06681	1·03955	1·01389	0·98967	0·96673
27	1·23616	1·19647	1·16010	1·12655	1·09540	1·06634	1·03911	1·01348	0·98928	0·96635
28	1·23546	1·19584	1·15952	1·12601	1·09490	1·06588	1·03867	1·01306	0·98888	0·96598
29	1·23477	1·19520	1·15894	1·12548	1·09440	1·06541	1·03823	1·01265	0·98849	0·96561
30	1·23408	1·19457	1·15836	1·12494	1·09390	1·06494	1·03779	1·01223	0·98810	0·96524
31	1·23339	1·19395	1·15778	1·12440	1·09341	1·06447	1·03735	1·01182	0·98771	0·96487
32	1·23271	1·19332	1·15721	1·12387	1·09291	1·06401	1·03691	1·01141	0·98732	0·96450
33	1·23202	1·19269	1·15663	1·12333	1·09241	1·06354	1·03647	1·01100	0·98693	0·96413
34	1·23133	1·19206	1·15605	1·12280	1·09191	1·06308	1·03604	1·01058	0·98654	0·96376
35	1·23065	1·19144	1·15548	1·12227	1·09142	1·06261	1·03560	1·01017	0·98615	0·96339
36	1·22997	1·19081	1·15490	1·12173	1·09092	1·06215	1·03516	1·00976	0·98576	0·96302
37	1·22928	1·19019	1·15433	1·12120	1·09042	1·06168	1·03473	1·00935	0·98537	0·96265
38	1·22860	1·18957	1·15375	1·12067	1·08993	1·06122	1·03429	1·00894	0·98498	0·96228
39	1·22792	1·18895	1·15318	1·12014	1·08943	1·06076	1·03386	1·00853	0·98459	0·96191
40	1·22724	1·18833	1·15261	1·11961	1·08894	1·06030	1·03342	1·00812	0·98421	0·96154
41	1·22657	1·18771	1·15204	1·11908	1·08845	1·05983	1·03299	1·00771	0·98382	0·96117
42	1·22589	1·18709	1·15147	1·11855	1·08796	1·05937	1·03256	1·00730	0·98343	0·96081
43	1·22521	1·18647	1·15090	1·11802	1·08746	1·05891	1·03212	1·00689	0·98304	0·96044
44	1·22454	1·18585	1·15033	1·11750	1·08697	1·05845	1·03169	1·00648	0·98266	0·96007
45	1·22386	1·18523	1·14976	1·11697	1·08648	1·05799	1·03126	1·00607	0·98227	0·95971
46	1·22319	1·18462	1·14919	1·11644	1·08599	1·05753	1·03083	1·00567	0·98189	0·95934
47	1·22252	1·18400	1·14863	1·11592	1·08550	1·05707	1·03039	1·00526	0·98150	0·95897
48	1·22185	1·18339	1·14806	1·11539	1·08501	1·05662	1·02996	1·00485	0·98111	0·95861
49	1·22118	1·18278	1·14750	1·11487	1·08452	1·05616	1·02953	1·00445	0·98073	0·95824
50	1·22051	1·18217	1·14693	1·11435	1·08403	1·05570	1·02910	1·00404	0·98035	0·95788
51	1·21984	1·18155	1·14637	1·11382	1·08355	1·05524	1·02867	1·00363	0·97996	0·95751
52	1·21918	1·18094	1·14581	1·11330	1·08306	1·05479	1·02824	1·00323	0·97958	0·95715
53	1·21851	1·18033	1·14524	1·11278	1·08257	1·05443	1·02781	1·00282	0·97919	0·95678
54	1·21785	1·17973	1·14468	1·11226	1·08209	1·05388	1·02739	1·00242	0·97881	0·95642
55	1·21718	1·17912	1·14412	1·11174	1·08160	1·05342	1·02696	1·00202	0·97843	0·95606
56	1·21652	1·17851	1·14356	1·11122	1·08112	1·05297	1·02653	1·00161	0·97805	0·95569
57	1·21586	1·17790	1·14300	1·11070	1·08063	1·05251	1·02610	1·00121	0·97766	0·95533
58	1·21520	1·17730	1·14244	1·11018	1·08015	1·05206	1·02568	1·00080	0·97728	0·95497
59	1·21454	1·17669	1·14189	1·10966	1·07966	1·05161	1·02525	1·00040	0·97690	0·95460
60	1·21388	1·17609	1·14133	1·10914	1·07918	1·05115	1·02482	1·00000	0·97652	0·95424

′	20°	21°	22°	23°	24°	25°	26°	27°	28°	29°
0	95424	93305	91285	89354	87506	85733	84030	82391	80811	79287
1	95388	93271	91252	89323	87476	85704	84002	82364	80786	79262
2	95352	93236	91219	89292	87446	85675	83974	82337	80760	79238
3	95316	93202	91186	89260	87416	85646	83946	82311	80734	79213
4	95280	93168	91154	89229	87386	85618	83919	82284	80708	79188
5	95244	93133	91121	89197	87356	85589	83891	82257	80682	79163
6	95208	93099	91088	89166	87326	85560	83863	82230	80657	79138
7	95172	93065	91055	89135	87296	85531	83835	82204	80631	79113
8	95136	93030	91023	89103	87266	85502	83808	82177	80605	79088
9	95100	92996	90990	89072	87236	85473	83780	82150	80579	79063
10	95064	92962	90957	89041	87206	85445	83752	82124	80554	79039
11	95028	92928	90925	89010	87176	85416	83725	82097	80528	79014
12	94992	92894	90892	88978	87146	85387	83697	82070	80502	78989
13	94956	92860	90859	88947	87116	85358	83670	82044	80477	78964
14	94921	92825	90827	88916	87086	85330	83642	82017	80451	78939
15	94885	92791	90794	88885	87056	85301	83614	81991	80425	78915
16	94849	92757	90762	88854	87026	85272	83587	81964	80400	78890
17	94813	92723	90729	88823	86996	85244	83559	81938	80374	78865
18	94778	92689	90697	88792	86967	85215	83532	81911	80349	78840
19	94742	92655	90664	88761	86937	85187	83504	81884	80323	78816
20	94706	92621	90632	88730	86907	85158	83477	81858	80297	78791
21	94671	92587	90599	88699	86877	85129	83449	81832	80272	78766
22	94635	92554	90567	88668	86848	85101	83422	81805	80246	78742
23	94600	92520	90535	88637	86818	85072	83394	81779	80221	78717
24	94564	92486	90502	88606	86788	85044	83367	81752	80195	78693
25	94529	92452	90470	88575	86759	85015	83339	81726	80170	78668
26	94493	92418	90438	88544	86729	84987	83312	81699	80144	78643
27	94458	92385	90406	88513	86699	84958	83285	81673	80119	78619
28	94423	92351	90373	88482	86670	84930	83257	81647	80094	78594
29	94387	92317	90341	88451	86640	84902	83230	81620	80068	78570
30	94352	92283	90309	88420	86611	84873	83203	81594	80043	78545
31	94317	92250	90277	88390	86581	84845	83175	81568	80017	78521
32	94281	92216	90245	88359	86552	84816	83148	81541	79992	78496
33	94246	92183	90213	88328	86522	84788	83121	81515	79967	78472
34	94211	92149	90181	88297	86493	84760	83094	81489	79941	78447
35	94176	92115	90148	88267	86463	84732	83066	81463	79916	78423
36	94141	92082	90116	88236	86434	84703	83039	81436	79891	78398
37	94105	92048	90084	88205	86404	84675	83012	81410	79865	78374
38	94070	92015	90052	88175	86375	84647	82985	81384	79840	78349
39	94035	91981	90020	88144	86346	84619	82958	81358	79815	78325
40	94000	91948	89988	88114	86316	84590	82930	81332	79790	78300
41	93965	91915	89957	88083	86287	84562	82903	81305	79764	78276
42	93930	91881	89925	88052	86258	84534	82876	81279	79739	78252
43	93895	91848	89893	88022	86228	84506	82849	81253	79714	78227
44	93860	91815	89861	87991	86199	84478	82822	81227	79689	78203
45	93825	91781	89829	87961	86170	84450	82795	81201	79663	78179
46	93791	91748	89797	87930	86140	84421	82768	81175	79638	78154
47	93756	91715	89766	87900	86111	84393	82741	81149	79613	78130
48	93721	91682	89734	87870	86082	84365	82714	81123	79588	78106
49	93686	91648	89702	87839	86053	84337	82687	81097	79563	78081
50	93651	91615	89670	87809	86024	84309	82660	81071	79538	78057
51	93617	91582	89639	87778	85995	84281	82633	81045	79513	78033
52	93582	91549	89607	87748	85965	84253	82606	81019	79488	78009
53	93547	91516	89575	87718	85936	84225	82579	80993	79463	77984
54	93513	91483	89544	87687	85907	84197	82552	80967	79437	77960
55	93478	91450	89512	87657	85878	84169	82525	80941	79412	77936
56	93443	91417	89481	87627	85849	84141	82498	80915	79387	77912
57	93409	91384	89449	87597	85820	84114	82471	80889	79362	77888
58	93374	91351	89417	87566	85791	84086	82445	80863	79337	77863
59	93340	91318	89386	87536	85762	84058	82418	80837	79312	77839
60	93305	91285	89354	87506	85733	84030	82391	80811	79287	77815

′	30°	31°	32°	33°	34°	35°	36°	37°	38°	39°
0	77815	76391	75012	73676	72379	71120	69897	68707	67549	66421
1	77791	76368	74990	73654	72358	71100	69877	68688	67530	66402
2	77767	76344	74967	73632	72337	71079	69857	68668	67511	66384
3	77743	76321	74944	73610	72316	71058	69837	68648	67492	66365
4	77719	76298	74922	73588	72294	71038	69817	68629	67473	66347
5	77695	76274	74899	73566	72273	71017	69797	68609	67454	66328
6	77671	76251	74877	73544	72252	70997	69777	68590	67435	66310
7	77647	76228	74854	73523	72231	70976	69756	68570	67416	66291
8	77623	76205	74832	73501	72209	70955	69736	68551	67397	66273
9	77599	76181	74809	73479	72188	70935	69716	68531	67378	66254
10	77575	76158	74787	73457	72167	70914	69696	68512	67359	66236
11	77551	76135	74764	73435	72146	70894	69676	68492	67340	66217
12	77527	76112	74742	73413	72125	70873	69656	68473	67321	66199
13	77503	76089	74719	73392	72103	70852	69636	68454	67302	66180
14	77479	76065	74697	73370	72082	70832	69616	68434	67283	66162
15	77455	76042	74674	73348	72061	70811	69596	68415	67264	66143
16	77431	76019	74652	73326	72040	70791	69576	68395	67245	66125
17	77407	75996	74629	73305	72019	70770	69557	68376	67226	66106
18	77383	75973	74607	73283	71998	70750	69537	68356	67207	66088
19	77359	75950	74585	73261	71977	70729	69517	68337	67188	66070
20	77335	75927	74562	73239	71956	70709	69497	68318	67170	66051
21	77311	75903	74540	73218	71935	70688	69477	68298	67151	66033
22	77288	75880	74517	73196	71914	70668	69457	68279	67132	66014
23	77264	75857	74495	73174	71892	70647	69437	68259	67113	65996
24	77240	75834	74473	73153	71871	70627	69417	68240	67094	65978
25	77216	75811	74450	73131	71850	70606	69397	68221	67075	65959
26	77192	75788	74428	73109	71829	70586	69377	68201	67056	65941
27	77169	75765	74406	73088	71808	70566	69358	68182	67038	65923
28	77145	75742	74383	73066	71787	70545	69338	68163	67019	65904
29	77121	75719	74361	73044	71766	70525	69318	68143	67000	65886
30	77097	75696	74339	73023	71745	70504	69298	68124	66981	65868
31	77974	75673	74317	73001	71724	70484	69278	68105	66962	65849
32	77050	75650	74294	72980	71703	70464	69258	68086	66944	65831
33	77026	75627	74272	72958	71682	70443	69239	68066	66925	65813
34	77002	75604	74250	72936	71662	70423	69219	68047	66906	65794
35	76979	75581	74228	72915	71641	70403	69199	68028	66887	65776
36	76955	75559	74205	72893	71620	70382	69179	68008	66869	65758
37	76931	75536	74183	72872	71599	70362	69159	67989	66850	65739
38	76908	75513	74161	72850	71578	70342	69140	67970	66831	65721
39	76884	75490	74139	72829	71557	70321	69120	67951	66812	65703
40	76861	75467	74117	72807	71536	70301	69100	67932	66794	65685
41	76837	75444	74095	72786	71515	70281	69080	67912	66775	65666
42	76813	75421	74072	72764	71494	70260	69061	67893	66756	65648
43	76790	75398	74050	72743	71473	70240	69041	67874	66737	65630
44	76766	75376	74028	72721	71453	70220	69021	67855	66719	65612
45	76743	75353	74006	72700	71432	70200	69002	67836	66700	65594
46	76719	75330	73984	72678	71411	70179	68982	67816	66681	65575
47	76696	75307	73962	72657	71390	70159	68962	67797	66663	65557
48	76672	75285	73940	72636	71369	70139	68942	67778	66644	65539
49	76649	75262	73918	72614	71349	70119	68923	67759	66625	65521
50	76625	75239	73896	72593	71328	70099	68903	67740	66607	65503
51	76602	75216	73874	72571	71307	70078	68884	67721	66588	65484
52	76578	75194	73852	72550	71286	70058	68864	67702	66570	65466
53	76555	75171	73830	72529	71265	70038	68844	67682	66551	65448
54	76531	75148	73808	72507	71245	70018	68825	67663	66532	65430
55	76508	75126	73786	72486	71224	69998	68805	67644	66514	65412
56	76485	75103	73764	72465	71203	69977	68785	67625	66495	65394
57	76461	75080	73742	72443	71183	69957	68766	67606	66477	65376
58	76438	75058	73720	72422	71162	69937	68746	67587	66458	65357
59	76414	75035	73698	72401	71141	69917	68727	67568	66439	65339
60	76391	75012	73676	72379	71120	69897	68707	67549	66421	65321

'	40°	41°	42°	43°	44°	45°	46°	47°	48°	49°
0	65321	64249	63202	62180	61182	60206	59251	58317	57403	56508
1	65303	64231	63185	62164	61166	60190	59236	58302	57388	56493
2	65285	64214	63168	62147	61149	60174	59220	58287	57373	56478
3	65267	64196	63151	62130	61133	60158	59204	58271	57358	56463
4	65249	64178	63133	62113	61116	60142	59189	58256	57343	56449
5	65231	64161	63116	62096	61100	60126	59173	58241	57328	56434
6	65213	64143	63099	62080	61083	60110	59157	58225	57313	56419
7	65195	64125	63082	62063	61067	60094	59141	58210	57298	56404
8	65177	64108	63065	62046	61051	60078	59126	58194	57283	56390
9	65159	64090	63047	62029	61034	60061	59110	58179	57268	56375
10	65141	64073	63030	62012	61018	60045	59094	58164	57253	56360
11	65123	64055	63013	61996	61001	60029	59079	58148	57238	56345
12	65105	64038	62996	61979	60985	60013	59063	58133	57223	56331
13	65087	64020	62979	61962	60969	59997	59047	58118	57208	56316
14	65069	64002	62962	61945	60952	59981	59032	58102	57193	56301
15	65051	63985	62945	61929	60936	59965	59016	58087	57178	56287
16	65033	63967	62927	61912	60920	59949	59000	58072	57163	56272
17	65015	63950	62910	61895	60903	59933	58985	58056	57148	56257
18	64997	63932	62893	61878	60887	59917	58969	58041	57133	56243
19	64979	63915	62876	61862	60871	59901	58954	58026	57118	56228
20	64961	63897	62859	61845	60854	59885	58938	58011	57103	56213
21	64943	63880	62842	61828	60838	59870	58922	57995	57088	56199
22	64925	63862	62825	61812	60822	59854	58907	57980	57073	56184
23	64907	63845	62808	61795	60805	59838	58891	57965	57058	56169
24	64889	63827	62791	61778	60789	59822	58875	57949	57043	56155
25	64871	63810	62774	61762	60773	59806	58860	57934	57028	56140
26	64853	63792	62757	61745	60756	59790	58844	57919	57013	56125
27	64835	63775	62739	61728	60740	59774	58829	57904	56998	56111
28	64818	63757	62722	61712	60724	59758	58813	57888	56983	56096
29	64800	63740	62705	61695	60708	59742	58798	57873	56968	56081
30	64782	63722	62688	61678	60691	59726	58782	57858	56953	56067
31	64764	63705	62671	61662	60675	59710	58766	57843	56938	56052
32	64746	63688	62654	61645	60659	59694	58751	57827	56923	56037
33	64728	63670	62637	61628	60642	59678	58735	57812	56908	56023
34	64710	63653	62620	61612	60626	59663	58720	57797	56893	56008
35	64692	63635	62603	61595	60610	59647	58704	57782	56879	55994
36	64675	63618	62586	61579	60594	59631	58689	57767	56864	55979
37	64657	63601	62569	61562	60578	59615	58673	57751	56849	55964
38	64639	63583	62552	61545	60561	59599	58658	57736	56834	55950
39	64621	63566	62535	61529	60545	59583	58642	57721	56819	55935
40	64603	63548	62518	61512	60529	59567	58627	57706	56804	55921
41	64586	63531	62501	61496	60513	59551	58611	57691	56789	55906
42	64568	63514	62484	61479	60496	59536	58596	57675	56774	55892
43	64550	63496	62468	61463	60480	59520	58580	57660	56759	55877
44	64532	63479	62451	61446	60464	59504	58565	57645	56745	55862
45	64514	63462	62434	61429	60448	59488	58549	57630	56730	55848
46	64497	63444	62417	61413	60432	59472	58534	57615	56715	55833
47	64479	63427	62400	61396	60416	59457	58518	57600	56700	55819
48	64461	63410	62383	61380	60399	59441	58503	57584	56685	55804
49	64443	63392	62366	61363	60383	59425	58487	57569	56670	55790
50	64426	63375	62349	61347	60367	59409	58472	57554	56656	55775
51	64408	63358	62332	61330	60351	59393	58456	57539	56641	55761
52	64390	63340	62315	61314	60335	59378	58441	57524	56626	55746
53	64373	63323	62298	61297	60319	59362	58425	57509	56611	55732
54	64355	63306	62282	61281	60303	59346	58410	57494	56596	55717
55	64337	63289	62265	61264	60286	59330	58395	57479	56582	55703
56	64320	63271	62248	61248	60270	59314	58379	57463	56567	55688
57	64302	63254	62231	61231	60254	59299	58364	57448	56552	55674
58	64284	63237	62214	61215	60238	59283	58348	57433	56537	55659
59	64267	63220	62197	61198	60222	59267	58333	57418	56522	55645
60	64249	63202	62180	61182	60206	59251	58317	57403	56508	55630

TERNARY PROPORTIONAL LOGARITHMS

′	50°	51°	52°	53°	54°	55°	56°	57°	58°	59°
0	55630	54770	53927	53100	52288	51491	50708	49940	49184	48442
1	55616	54756	53913	53086	52274	51478	50696	49927	49172	48430
2	55601	54742	53899	53072	52261	51465	50683	49914	49159	48418
3	55587	54728	53885	53059	52248	51452	50670	49902	49147	48405
4	55572	54714	53871	53045	52234	51438	50657	49889	49135	48393
5	55558	54699	53857	53031	52221	51425	50644	49876	49122	48381
6	55543	54685	53843	53018	52208	51412	50631	49864	49110	48369
7	55529	54671	53830	53004	52194	51399	50618	49851	49097	48356
8	55515	54657	53816	52991	52181	51386	50605	49838	49085	48344
9	55500	54643	53802	52977	52167	51373	50592	49826	49072	48332
10	55486	54629	53788	52963	52154	51360	50579	49813	49060	48320
11	55471	54614	53774	52950	52141	51346	50566	49800	49047	48307
12	55457	54600	53760	52936	52127	51333	50554	49788	49035	48295
13	55442	54586	53746	52922	52114	51320	50541	49775	49023	48283
14	55428	54572	53732	52909	52101	51307	50528	49762	49010	48271
15	55414	54558	53719	52895	52087	51294	50515	49750	48998	48258
16	55399	54544	53705	52882	52074	51281	50502	49737	48985	48246
17	55385	54530	53691	52868	52061	51268	50489	49724	48973	48234
18	55370	54516	53677	52855	52047	51255	50476	49712	48960	48222
19	55356	54501	53663	52841	52034	51242	50464	49699	48948	48210
20	55342	54487	53649	52827	52021	51229	50451	49687	48936	48197
21	55327	54473	53636	52814	52007	51215	50438	49674	48923	48185
22	55313	54459	53622	52800	51994	51202	50425	49661	48911	48173
23	55299	54445	53608	52787	51981	51189	50412	49649	48898	48161
24	55284	54431	53594	52773	51967	51176	50399	49636	48886	48149
25	55270	54417	53580	52760	51954	51163	50387	49623	48874	48136
26	55255	54403	53567	52746	51941	51150	50374	49611	48861	48124
27	55241	54389	53553	52732	51927	51137	50361	49598	48849	48112
28	55227	54375	53539	52719	51914	51124	50348	49586	48836	48100
29	55212	54361	53525	52705	51901	51111	50335	49573	48824	48088
30	55198	54347	53511	52692	51888	51098	50322	49560	48812	48076
31	55184	54332	53498	52678	51874	51085	50310	49548	48799	48063
32	55169	54318	53484	52665	51861	51072	50297	49535	48787	48051
33	55155	54304	53470	52651	51848	51059	50284	49523	48775	48039
34	55141	54290	53456	52638	51835	51046	50271	49510	48762	48027
35	55127	54276	53442	52624	51821	51033	50258	49498	48750	48015
36	55112	54262	53429	52611	51808	51020	50246	49485	48737	48003
37	55098	54248	53415	52597	51795	51007	50233	49472	48725	47990
38	55084	54234	53401	52584	51781	50994	50220	49460	48713	47978
39	55069	54220	53387	52570	51768	50981	50207	49447	48700	47966
40	55055	54206	53374	52557	51755	50968	50194	49435	48688	47954
41	55041	54192	53360	52543	51742	50955	50182	49422	48676	47942
42	55026	54178	53346	52530	51729	50942	50169	49410	48663	47930
43	55012	54164	53332	52516	51715	50929	50156	49397	48651	47918
44	54998	54150	53319	52503	51702	50916	50143	49385	48639	47906
45	54984	54136	53305	52489	51689	50903	50131	49372	48626	47893
46	54969	54122	53291	52476	51676	50890	50118	49360	48614	47881
47	54955	54108	53278	52462	51662	50877	50105	49347	48602	47869
48	54941	54094	53264	52449	51649	50864	50092	49334	48590	47857
49	54927	54080	53250	52436	51636	50851	50080	49322	48577	47845
50	54912	54066	53236	52422	51623	50838	50067	49309	48565	47833
51	54898	54052	53223	52409	51610	50825	50054	49297	48553	47821
52	54884	54038	53209	52395	51596	50812	50041	49284	48540	47809
53	54870	54024	53195	52382	51583	50799	50029	49272	48528	47797
54	54855	54011	53182	52368	51570	50786	50016	49259	48516	47785
55	54841	53997	53168	52355	51557	50773	50003	49247	48503	47772
56	54827	53983	53154	52342	51544	50760	49991	49234	48491	47760
57	54813	53969	53141	52328	51530	50747	49978	49222	48479	47748
58	54799	53955	53127	52315	51517	50734	49965	49209	48467	47736
59	54784	53941	53113	52301	51504	50721	49952	49197	48454	47724
60	54770	53927	53100	52288	51491	50708	49940	49184	48442	47712

′	60°	61°	62°	63°	64°	65°	66°	67°	68°	69°	70°	71°
0	47712	46994	46288	45593	44909	44236	43573	42920	42276	41642	41017	40401
1	47700	46982	46276	45582	44898	44225	43562	42909	42266	41632	41007	40391
2	47688	46971	46265	45570	44887	44214	43551	42898	42255	41621	40997	40381
3	47676	46959	46253	45559	44875	44203	43540	42887	42244	41611	40986	40371
4	47664	46947	46241	45547	44864	44191	43529	42877	42234	41600	40976	40361
5	47652	46935	46230	45536	44853	44180	43518	42866	42223	41590	40966	40350
6	47640	46923	46218	45524	44841	44169	43507	42855	42213	41579	40955	40340
7	47628	46911	46206	45513	44830	44158	43496	42844	42202	41569	40945	40330
8	47616	46899	46195	45501	44819	44147	43485	42833	42191	41559	40935	40320
9	47604	46888	46183	45490	44808	44136	43474	42823	42181	41548	40924	40310
10	47592	46876	46171	45478	44796	44125	43463	42812	42170	41538	40914	40300
11	47580	46864	46160	45467	44785	44114	43452	42801	42159	41527	40904	40289
12	47568	46852	46148	45456	44774	44102	43441	42790	42149	41517	40894	40279
13	47556	46840	46137	45444	44762	44091	43431	42780	42138	41506	40883	40269
14	47544	46828	46125	45433	44751	44080	43420	42769	42128	41496	40873	40259
15	47532	46817	46113	45421	44740	44069	43409	42758	42117	41485	40863	40249
16	47520	46805	46102	45410	44729	44058	43398	42747	42106	41475	40852	40239
17	47508	46793	46090	45398	44717	44047	43387	42737	42096	41464	40842	40228
18	47496	46781	46078	45387	44706	44036	43376	42726	42085	41454	40832	40218
19	47484	46769	46067	45375	44695	44025	43365	42715	42075	41443	40821	40208
20	47472	46758	46055	45364	44684	44014	43354	42704	42064	41433	40811	40198
21	47460	46746	46044	45353	44672	44003	43343	42693	42053	41423	40801	40188
22	47448	46734	46032	45341	44661	43992	43332	42683	42043	41412	40791	40178
23	47436	46722	46020	45330	44650	43981	43321	42672	42032	41402	40780	40168
24	47424	46710	46009	45318	44639	43969	43310	42661	42022	41391	40770	40157
25	47412	46699	45997	45307	44627	43958	43300	42651	42011	41381	40760	40147
26	47400	46687	45986	45295	44616	43947	43289	42640	42000	41370	40749	40137
27	47388	46675	45974	45284	44605	43936	43278	42629	41990	41360	40739	40127
28	47376	46663	45962	45273	44594	43925	43267	42618	41979	41350	40729	40117
29	47364	46652	45951	45261	44583	43914	43256	42608	41969	41339	40719	40107
30	47352	46640	45939	45250	44571	43903	43245	42597	41958	41329	40708	40097
31	47340	46628	45928	45238	44560	43892	43234	42586	41948	41318	40698	40087
32	47328	46616	45916	45227	44549	43881	43223	42575	41937	41308	40688	40076
33	47316	46604	45905	45216	44538	43870	43212	42565	41927	41298	40678	40066
34	47304	46593	45893	45204	44526	43859	43202	42554	41916	41287	40667	40056
35	47292	46581	45881	45193	44515	43848	43191	42543	41905	41277	40657	40046
36	47280	46569	45870	45182	44504	43837	43180	42533	41895	41266	40647	40036
37	47268	46557	45858	45170	44493	43826	43169	42522	41884	41256	40637	40026
38	47256	46546	45847	45159	44482	43815	43158	42511	41874	41246	40626	40016
39	47244	46534	45835	45147	44470	43804	43147	42500	41863	41235	40616	40006
40	47232	46522	45824	45136	44459	43793	43136	42490	41853	41225	40606	39996
41	47220	46510	45812	45125	44448	43782	43126	42479	41842	41214	40596	39985
42	47208	46499	45800	45113	44437	43771	43115	42468	41832	41204	40585	39975
43	47196	46487	45789	45102	44426	43760	43104	42458	41821	41194	40575	39965
44	47185	46475	45777	45091	44414	43749	43093	42447	41811	41183	40565	39955
45	47173	46464	45766	45079	44403	43738	43082	42436	41800	41173	40555	39945
46	47161	46452	45754	45068	44392	43727	43071	42426	41789	41162	40544	39935
47	47149	46440	45743	45057	44381	43716	43060	42415	41779	41152	40534	39925
48	47137	46428	45731	45045	44370	43705	43050	42404	41768	41142	40524	39915
49	47125	46417	45720	45034	44359	43694	43039	42394	41758	41131	40514	39905
50	47113	46405	45708	45022	44347	43683	43028	42383	41747	41121	40503	39895
51	47101	46393	45697	45011	44336	43672	43017	42372	41737	41111	40493	39885
52	47089	46382	45685	45000	44325	43661	43006	42362	41726	41100	40483	39874
53	47077	46370	45674	44988	44314	43650	42995	42351	41716	41090	40473	39864
54	47066	46358	45662	44977	44303	43639	42985	42340	41705	41080	40463	39854
55	47054	46346	45651	44966	44292	43628	42974	42330	41695	41069	40452	39844
56	47042	46335	45639	44955	44280	43617	42963	42319	41684	41059	40442	39834
57	47030	46323	45628	44943	44269	43606	42952	42308	41674	41048	40432	39824
58	47018	46311	45616	44932	44258	43595	42941	42298	41663	41038	40422	39814
59	47006	46300	45605	44921	44247	43584	42931	42287	41653	41028	40412	39804
60	46994	46288	45593	44909	44236	43573	42920	42276	41642	41017	40401	39794

′	72°	73°	74°	75°	76°	77°	78°	79°	80°	81°	82°	83°
0	39794	39195	38604	38021	37446	36878	36318	35765	35218	34679	34146	33619
1	39784	39185	38594	38011	37436	36869	36309	35755	35209	34670	34137	33611
2	39774	39175	38585	38002	37427	36859	36299	35746	35200	34661	34128	33602
3	39764	39165	38575	37992	37417	36850	36290	35737	35191	34652	34119	33593
4	39754	39155	38565	37983	37408	36841	36281	35728	35182	34643	34111	33585
5	39744	39145	38555	37973	37398	36831	36271	35719	35173	34634	34102	33576
6	39734	39136	38545	37963	37389	36822	36262	35710	35164	34625	34093	33567
7	39724	39126	38536	37954	37379	36812	36253	35700	35155	34616	34084	33558
8	39714	39116	38526	37944	37370	36803	36244	35691	35146	34607	34075	33550
9	39704	39106	38516	37934	37360	36794	36234	35682	35137	34598	34066	33541
10	39694	39096	38506	37925	37351	36784	36225	35673	35128	34589	34058	33532
11	39684	39086	38497	37915	37341	36775	36216	35664	35119	34581	34049	33524
12	39674	39076	38487	37905	37332	36766	36207	35655	35110	34572	34040	33515
13	39664	39066	38477	37896	37322	36756	36197	35646	35101	34563	34031	33506
14	39653	39056	38467	37886	37313	36747	36188	35636	35092	34554	34022	33498
15	39643	39046	38458	37877	37303	36737	36179	35627	35083	34545	34014	33489
16	39633	39037	38448	37867	37294	36728	36170	35618	35074	34536	34005	33480
17	39623	39027	38438	37857	37284	36719	36160	35609	35065	34527	33996	33471
18	39613	39017	38428	37848	37275	36709	36151	35600	35056	34518	33987	33463
19	39603	39007	38419	37838	37265	36700	36142	35591	35047	34509	33978	33454
20	39593	38997	38409	37829	37256	36691	36133	35582	35038	34500	33970	33445
21	39583	38987	38399	37819	37246	36681	36123	35573	35029	34491	33961	33437
22	39573	38977	38389	37809	37237	36672	36114	35563	35020	34483	33952	33428
23	39563	38968	38380	37800	37227	36663	36105	35554	35011	34474	33943	33419
24	39553	38958	38370	37790	37218	36653	36096	35545	35002	34465	33935	33411
25	39543	38948	38360	37781	37208	36644	36086	35536	34993	34456	33926	33402
26	39533	38938	38351	37771	37199	36634	36077	35527	34984	34447	33917	33393
27	39523	38928	38341	37761	37189	36625	36068	35518	34975	34438	33908	33385
28	39513	38918	38331	37752	37180	36616	36059	35509	34966	34429	33899	33376
29	39503	38908	38321	37742	37171	36606	36050	35500	34957	34420	33891	33367
30	39493	38899	38312	37733	37161	36597	36040	35491	34948	34411	33882	33359
31	39483	38889	38302	37723	37152	36588	36031	35481	34939	34403	33873	33350
32	39473	38879	38292	37713	37142	36578	36022	35472	34930	34394	33864	33341
33	39464	38869	38282	37704	37133	36569	36013	35463	34921	34385	33856	33333
34	39454	38859	38273	37694	37123	36560	36003	35454	34912	34376	33847	33324
35	39444	38849	38263	37685	37114	36550	35994	35445	34903	34367	33838	33315
36	39434	38839	38253	37675	37104	36541	35985	35436	34894	34358	33829	33307
37	39424	38830	38244	37665	37095	36532	35976	35427	34885	34349	33820	33298
38	39414	38820	38234	37656	37085	36522	35967	35418	34876	34340	33812	33289
39	39404	38810	38224	37646	37076	36513	35957	35409	34867	34332	33803	33281
40	39394	38800	38215	37637	37067	36504	35948	35400	34858	34323	33794	33272
41	39384	38790	38205	37627	37057	36494	35939	35391	34849	34314	33785	33263
42	39374	38781	38195	37618	37048	36485	35930	35381	34840	34305	33777	33255
43	39364	38771	38186	37608	37038	36476	35921	35372	34831	34296	33768	33246
44	39354	38761	38176	37599	37029	36467	35911	35363	34822	34287	33759	33237
45	39344	38751	38166	37589	37019	36457	35902	35354	34813	34278	33750	33229
46	39334	38741	38156	37579	37010	36448	35893	35345	34804	34270	33742	33220
47	39324	38731	38147	37570	37001	36439	35884	35336	34795	34261	33733	33211
48	39314	38722	38137	37560	36991	36429	35875	35327	34786	34252	33724	33203
49	39304	38712	38127	37551	36982	36420	35865	35318	34777	34243	33715	33194
50	39294	38702	38118	37541	36972	36411	35856	35309	34768	34234	33707	33186
51	39284	38692	38108	37532	36963	36401	35847	35300	34759	34225	33698	33177
52	39274	38682	38098	37522	36953	36392	35838	35291	34750	34217	33689	33168
53	39264	38673	38089	37513	36944	36383	35829	35282	34741	34208	33681	33160
54	39254	38663	38079	37503	36935	36374	35820	35273	34732	34199	33672	33151
55	39245	38653	38069	37494	36925	36364	35810	35264	34723	34190	33663	33142
56	39235	38643	38060	37484	36916	36355	35801	35254	34715	34181	33654	33134
57	39225	38633	38050	37474	36906	36346	35792	35245	34706	34172	33646	33125
58	39215	38624	38040	37465	36897	36336	35783	35236	34697	34164	33637	33117
59	39205	38614	38031	37455	36888	36327	35774	35227	34688	34155	33628	33108
60	39195	38604	38021	37446	36878	36318	35765	35218	34679	34146	33619	33099

′	84°	85°	86°	87°	88°	89°	90°	91°	92°	93°	94°	95°
0	33099	32585	32077	31575	31079	30588	30103	29623	29148	28679	28214	27755
1	33091	32577	32069	31567	31071	30580	30095	29615	29141	28671	28207	27747
2	33082	32568	32061	31559	31063	30572	30087	29607	29133	28663	28199	27740
3	33073	32560	32052	31550	31054	30564	30079	29599	29125	28656	28191	27732
4	33065	32551	32044	31542	31046	30556	30071	29591	29117	28648	28184	27724
5	33056	32543	32035	31534	31038	30548	30063	29583	29109	28640	28176	27717
6	33048	32534	32027	31525	31030	30539	30055	29575	29101	28632	28168	27709
7	33039	32526	32019	31517	31021	30531	30047	29567	29093	28625	28161	27702
8	33030	32517	32010	31509	31013	30523	30039	29560	29086	28617	28153	27694
9	33022	32509	32002	31501	31005	30515	30031	29552	29078	28609	28145	27686
10	33013	32500	31993	31492	30997	30507	30023	29544	29070	28601	28138	27679
11	33005	32492	31985	31484	30989	30499	30015	29536	29062	28593	28130	27671
12	32996	32483	31977	31476	30980	30491	30007	29528	29054	28586	28122	27664
13	32987	32475	31968	31467	30972	30483	29999	29520	29046	28578	28114	27656
14	32979	32466	31960	31459	30964	30475	29991	29512	29038	28570	28107	27648
15	32970	32458	31951	31451	30956	30466	29983	29504	29031	28562	28099	27641
16	32962	32449	31943	31442	30948	30458	29975	29496	29023	28555	28091	27633
17	32953	32441	31935	31434	30939	30450	29967	29488	29015	28547	28084	27626
18	32944	32432	31926	31426	30931	30442	29958	29480	29007	28539	28076	27618
19	32936	32424	31918	31418	30923	30434	29950	29472	28999	28531	28068	27610
20	32927	32415	31909	31409	30915	30426	29942	29464	28991	28524	28061	27603
21	32919	32407	31901	31401	30907	30418	29934	29456	28984	28516	28053	27595
22	32910	32398	31893	31393	30898	30410	29926	29448	28976	28508	28045	27588
23	32902	32390	31884	31384	30890	30302	29918	29441	28968	28500	28038	27580
24	32893	32381	31876	31376	30882	30393	29910	29433	28960	28493	28030	27572
25	32884	32373	31867	31368	30874	30385	29902	29425	28952	28485	28022	27565
26	32876	32365	31859	31360	30866	30377	29894	29417	28944	28477	28015	27557
27	32867	32356	31851	31351	30857	30369	29886	29409	28937	28469	28007	27550
28	32859	32348	31842	31343	30849	30361	29878	29401	28929	28462	27999	27542
29	32850	32339	31834	31335	30841	30353	29870	29393	28921	28454	27992	27534
30	32842	32331	31826	31326	30833	30345	29862	29385	28913	28446	27984	27527
31	32833	32322	31817	31318	30825	30337	29854	29377	28905	28438	27976	27519
32	32824	32314	31809	31310	30817	30329	29846	29369	28897	28431	27969	27512
33	32816	32305	31801	31302	30808	30321	29838	29361	28890	28423	27961	27504
34	32807	32297	31792	31293	30800	30313	29830	29354	28882	28415	27953	27497
35	32799	32288	31784	31285	30792	30305	29822	29346	28874	28407	27946	27489
36	32790	32280	31775	31277	30784	30296	29814	29338	28866	28400	27938	27481
37	32782	32271	31767	31269	30776	30?88	29806	29330	28858	28392	27930	27474
38	32773	32263	31759	31260	30768	30280	29798	29322	28851	28384	27923	27466
39	32765	32255	31750	31252	30759	30272	29790	29314	28843	28376	27915	27459
40	32756	32246	31742	31244	30751	30264	29782	29306	28835	28369	27908	27451
41	32747	32238	31734	31236	30743	30256	29775	29298	28827	28361	27900	27444
42	32739	32229	31725	31227	30735	30248	29767	29290	28819	28353	27892	27436
43	32730	32221	31717	31219	30727	30240	29759	29282	28811	28346	27885	27429
44	32722	32212	31709	31211	30719	30232	29751	29275	28804	28338	27877	27421
45	32713	32204	31700	31203	30710	30224	29743	29267	28796	28330	27869	27413
46	32705	32195	31692	31194	30702	30216	29735	29259	28788	28322	27862	27406
47	32696	32187	31684	31186	30694	30208	29727	29251	28780	28315	27854	27398
48	32688	32179	31675	31178	30686	30200	29719	29243	28772	28307	27846	27391
49	32679	32170	31667	31170	30678	30192	29711	29235	28765	28299	27839	27383
50	32671	32162	31659	31161	30670	30183	29703	29227	28757	28292	27831	27376
51	32662	32153	31650	31153	30662	30175	29695	29219	28749	28284	27824	27368
52	32654	32145	31642	31145	30653	30167	29687	29211	28741	28276	27816	27360
53	32645	32136	31634	31137	30645	30159	29679	29204	28733	28268	27808	27353
54	32636	32128	31625	31128	30637	30151	29671	29196	28726	28261	27801	27345
55	32628	32120	31617	31120	30629	30143	29663	29188	28718	28253	27793	27338
56	32619	32111	31609	31112	30621	30135	29655	29180	28710	28245	27785	27330
57	32611	32103	31600	31104	30613	30127	29647	29172	28702	28238	27778	27323
58	32602	32094	31592	31095	30605	30119	29639	29164	28695	28230	27770	27315
59	32594	32086	31584	31087	30596	30111	29631	29156	28687	28222	27763	27308
60	32585	32077	31575	31079	30588	30103	29623	29148	28679	28214	27755	27300

′	96°	97°	98°	99°	100°	101°	102°	103°	104°	105°	106°	107°
0	27300	26850	26405	25964	25527	25095	24667	24244	23824	23408	22997	22589
1	27293	26843	26397	25956	25520	25088	24660	24237	23817	23401	22990	22582
2	27285	26835	26390	25949	25513	25081	24653	24229	23810	23395	22983	22575
3	27278	26828	26382	25942	25506	25074	24646	24222	23803	23388	22976	22569
4	27270	26820	26375	25934	25498	25066	24639	24215	23796	23381	22969	22562
5	27262	26813	26368	25927	25491	25059	24632	24208	23789	23374	22963	22555
6	27255	26805	26360	25920	25484	25052	24625	24201	23782	23367	22956	22548
7	27247	26798	26353	25913	25477	25045	24618	24194	23775	23360	22949	22542
8	27240	26790	26346	25905	25469	25038	24610	24187	23768	23353	22942	22535
9	27232	26783	26338	25898	25462	25031	24603	24180	23761	23346	22935	22528
10	27225	26776	26331	25891	25455	25024	24596	24173	23754	23339	22928	22521
11	27217	26768	26323	25883	25448	25016	24589	24166	23747	23333	22922	22515
12	27210	26761	26316	25876	25440	25009	24582	24159	23740	23326	22915	22508
13	27202	26753	26309	25869	25433	25002	24575	24152	23734	23319	22908	22501
14	27195	26746	26301	25861	25426	24995	24568	24145	23727	23312	22901	22494
15	27187	26738	26294	25854	25419	24988	24561	24138	23720	23305	22894	22488
16	27180	26731	26287	25847	25412	24981	24554	24131	23713	23298	22888	22481
17	27172	26723	26279	25840	25404	24973	24547	24124	23706	23291	22881	22474
18	27165	26716	26272	25832	25397	24966	24540	24117	23699	23284	22874	22467
19	27157	26709	26265	25825	25390	24959	24533	24110	23692	23278	22867	22461
20	27150	26701	26257	25818	25383	24952	24526	24103	23685	23271	22860	22454
21	27142	26694	26250	25810	25376	24945	24518	24096	23678	23264	22854	22447
22	27135	26686	26242	25803	25368	24938	24511	24089	23671	23257	22847	22440
23	27127	26679	26235	25796	25361	24931	24504	24082	23664	23250	22840	22434
24	27120	26671	26228	25789	25354	24923	24497	24075	23657	23243	22833	22427
25	27112	26664	26220	25781	25347	24916	24490	24068	23650	23236	22826	22420
26	27105	26656	26213	25774	25339	24909	24483	24061	23643	23229	22819	22413
27	27097	26649	26206	25767	25332	24902	24476	24054	23636	23223	22813	22407
28	27090	26642	26198	25759	25325	24895	24469	24047	23629	23216	22806	22400
29	27082	26634	26191	25752	25318	24888	24462	24040	23623	23209	22799	22393
30	27075	26627	26184	25745	25311	24881	24455	24033	23616	23202	22792	22386
31	27067	26619	26176	25738	25303	24874	24448	24026	23609	23195	22785	22380
32	27060	26612	26169	25730	25296	24866	24441	24019	23602	23188	22779	22373
33	27052	26605	26162	25723	25289	24859	24434	24012	23595	23181	22772	22366
34	27045	26597	26154	25716	25282	24852	24427	24005	23588	23175	22765	22359
35	27037	26590	26147	25709	25275	24845	24420	23998	23581	23168	22758	22353
36	27030	26582	26140	25701	25267	24838	24413	23991	23574	23161	22752	22346
37	27022	26575	26132	25694	25260	24831	24405	23984	23567	23154	22745	22339
38	27015	26567	26125	25687	25253	24824	24398	23977	23560	23147	22738	22333
39	27007	26560	26118	25680	25246	24817	24391	23970	23553	23140	22731	22326
40	27000	26553	26110	25672	25239	24809	24384	23963	23546	23133	22724	22319
41	26992	26545	26103	25665	25231	24802	24377	23956	23539	23127	22718	22312
42	26985	26538	26096	25658	25224	24795	24370	23949	23533	23120	22711	22306
43	26977	26530	26088	25650	25217	24788	24363	23942	23526	23113	22704	22299
44	26970	26523	26081	25643	25210	24781	24356	23935	23519	23106	22697	22292
45	26962	26516	26074	25636	25203	24774	24349	23928	23512	23099	22690	22286
46	26955	26508	26066	25629	25196	24767	24342	23921	23505	23092	22684	22279
47	26947	26501	26059	25621	25188	24760	24335	23914	23498	23086	22677	22272
48	26940	26493	26052	25614	25181	24752	24328	23908	23491	23079	22670	22265
49	26932	26486	26044	25607	25174	24745	24321	23901	23484	23072	22663	22259
50	26925	26479	26037	25600	25167	24738	24314	23894	23477	23065	22657	22252
51	26917	26471	26030	25592	25160	24731	24307	23887	23470	23058	22650	22245
52	26910	26464	26022	25585	25152	24724	24300	23880	23464	23051	22643	22239
53	26902	26457	26015	25578	25145	24717	24293	23873	23457	23044	22636	22232
54	26895	26449	26008	25571	25138	24710	24286	23866	23450	23038	22629	22225
55	26887	26442	26000	25563	25131	24703	24279	23859	23443	23031	22623	22218
56	26880	26434	25993	25556	25124	24696	24272	23852	23436	23024	22616	22212
57	26872	26427	25986	25549	25117	24689	24265	23845	23429	23017	22609	22205
58	26865	26419	25978	25542	25109	24681	24258	23838	23422	23010	22602	22198
59	26858	26412	25971	25534	25102	24674	24251	23831	23415	23004	22596	22192
60	26850	26405	25964	25527	25095	24667	24244	23824	23408	22997	22589	22185

′	108°	109°	110°	111°	112°	113°	114°	115°	116°	117°	118°	119°
0	22185	21785	21388	20995	20605	20219	19837	19457	19081	18709	18339	17973
1	22178	21778	21381	20988	20599	20213	19830	19451	19075	18702	18333	17966
2	22171	21771	21375	20982	20593	20207	19824	19445	19069	18696	18327	17960
3	22165	21765	21368	20975	20586	20200	19818	19439	19063	18690	18321	17954
4	22158	21758	21362	20969	20580	20194	19811	19432	19056	18684	18315	17948
5	22151	21751	21355	20962	20573	20187	19805	19426	19050	18678	18308	17942
6	22145	21745	21349	20956	20567	20181	19799	19420	19044	18672	18302	17936
7	22138	21738	21342	20949	20560	20175	19792	19413	19038	18665	18296	17930
8	22131	21732	21335	20943	20554	20168	19786	19407	19032	18659	18290	17924
9	22125	21725	21329	20936	20547	20162	19780	19401	19025	18653	18284	17918
10	22118	21718	21322	20930	20541	20155	19773	19395	19019	18647	18278	17912
11	22111	21712	21316	20923	20534	20149	19767	19388	19013	18641	18272	17906
12	22105	21705	21309	20917	20528	20143	19761	19382	19007	18634	18266	17900
13	22098	21698	21303	20910	20522	20136	19754	19376	19000	18628	18259	17894
14	22091	21692	21296	20904	20515	20130	19748	19369	18994	18622	18253	17887
15	22084	21685	21289	20897	20509	20123	19742	19363	18988	18616	18247	17881
16	22078	21678	21283	20891	20502	20117	19735	19357	18982	18610	18241	17875
17	22071	21672	21276	20884	20496	20111	19729	19351	18976	18604	18235	17869
18	22064	21665	21270	20878	20489	20104	19723	19344	18969	18597	18229	17863
19	22058	21659	21263	20871	20483	20098	19716	19338	18963	18591	18223	17857
20	22051	21652	21257	20865	20476	20091	19710	19332	18957	18585	18217	17851
21	22044	21645	21250	20858	20470	20085	19704	19325	18951	18579	18210	17845
22	22038	21639	21243	20852	20464	20079	19697	19319	18944	18573	18204	17839
23	22031	21632	21237	20845	20457	20072	19691	19313	18938	18567	18198	17833
24	22024	21626	21230	20839	20451	20066	19685	19307	18932	18560	18192	17827
25	22018	21619	21224	20832	20444	20060	19678	19300	18926	18554	18186	17821
26	22011	21612	21217	20826	20438	20053	19672	19294	18920	18548	18180	17815
27	22004	21606	21211	20819	20431	20047	19666	19288	18913	18542	18174	17809
28	21998	21599	21204	20813	20425	20040	19659	19282	18907	18536	18168	17803
29	21991	21592	21198	20806	20418	20034	19653	19275	18901	18530	18162	17797
30	21984	21586	21191	20800	20412	20028	19647	19269	18895	18523	18155	17790
31	21978	21579	21184	20793	20406	20021	19640	19263	18888	18517	18149	17784
32	21971	21573	21178	20787	20399	20015	19634	19257	18882	18511	18143	17778
33	21964	21566	21171	20780	20393	20009	19628	19250	18876	18505	18137	17772
34	21958	21559	21165	20774	20386	20002	19621	19244	18870	18499	18131	17766
35	21951	21553	21158	20767	20380	19996	19615	19238	18864	18493	18125	17760
36	21944	21546	21152	20761	20373	19989	19609	19231	18857	18487	18119	17754
37	21938	21540	21145	20754	20367	19983	19602	19225	18851	18480	18113	17748
38	21931	21533	21139	20748	20361	19977	19596	19219	18845	18474	18107	17742
39	21924	21526	21132	20741	20354	19970	19590	19213	18839	18468	18100	17736
40	21918	21520	21126	20735	20348	19964	19584	19206	18833	18462	18094	17730
41	21911	21513	21119	20728	20341	19958	19577	19200	18826	18456	18088	17724
42	21904	21507	21112	20722	20335	19951	19571	19194	18820	18450	18082	17718
43	21898	21500	21106	20715	20328	19945	19565	19188	18814	18443	18076	17712
44	21891	21493	21099	20709	20322	19938	19558	19181	18808	18437	18070	17706
45	21884	21487	21093	20702	20316	19932	19552	19175	18802	18431	18064	17700
46	21878	21480	21086	20696	20309	19926	19546	19169	18795	18425	18058	17694
47	21871	21474	21080	20690	20303	19919	19539	19163	18789	18419	18052	17688
48	21864	21467	21073	20683	20296	19913	19533	19156	18783	18413	18046	17682
49	21858	21460	21067	20676	20290	19907	19527	19150	18777	18407	18040	17676
50	21851	21454	21060	20670	20284	19900	19520	19144	18771	18400	18033	17669
51	21844	21447	21054	20664	20277	19894	19514	19138	18764	18394	18027	17663
52	21838	21441	21047	20657	20271	19888	19508	19131	18758	18388	18021	17657
53	21831	21434	21041	20651	20264	19881	19502	19125	18752	18382	18015	17651
54	21824	21427	21034	20644	20258	19875	19495	19119	18746	18376	18009	17645
55	21818	21421	21028	20638	20251	19869	19489	19113	18740	18370	18003	17639
56	21811	21414	21021	20631	20245	19862	19483	19106	18733	18364	17997	17633
57	21805	21408	21015	20625	20239	19856	19476	19100	18727	18357	17991	17627
58	21798	21401	21008	20618	20222	19849	19470	19094	18721	18351	17985	17621
59	21791	21395	21001	20612	20226	19843	19464	19088	18715	18345	17979	17615
60	21785	21388	20995	20605	20219	19837	19457	19081	18709	18339	17973	17609

′	120°	121°	122°	123°	124°	125°	126°	127°	128°	129°	130°	131°
0	17609	17249	16891	16537	16185	15836	15490	15147	14806	14468	14133	13800
1	17603	17243	16885	16531	16179	15830	15484	15141	14801	14463	14127	13795
2	17597	17237	16879	16525	16173	15825	15479	15135	14795	14457	14122	13789
3	17591	17231	16873	16519	16168	15819	15473	15130	14789	14451	14116	13784
4	17585	17225	16868	16513	16162	15813	15467	15124	14784	14446	14111	13778
5	17579	17219	16862	16507	16156	15807	15461	15118	14778	14440	14105	13773
6	17573	17213	16856	16501	16150	15802	15456	15113	14772	14435	14100	13767
7	17567	17207	16850	16496	16144	15796	15450	15107	14767	14429	14094	13761
8	17561	17201	16844	16490	16138	15790	15444	15101	14761	14423	14088	13756
9	17555	17195	16838	16484	16133	15784	15439	15096	14755	14418	14083	13750
10	17549	17189	16832	16478	16127	15778	15433	15090	14750	14412	14077	13745
11	17543	17183	16826	16472	16121	15773	15427	15084	14744	14407	14072	13739
12	17537	17177	16820	16466	16115	15767	15421	15079	14738	14401	14066	13734
13	17531	17171	16814	16460	16109	15761	15416	15073	14733	14395	14061	13728
14	17525	17165	16808	16454	16103	15755	15410	15067	14727	14390	14055	13723
15	17519	17159	16802	16449	16098	15749	15404	15061	14722	14384	14049	13717
16	17513	17153	16796	16443	16092	15744	15398	15056	14716	14379	14044	13712
17	17507	17147	16791	16437	16086	15738	15393	15050	14710	14373	14038	13706
18	17501	17141	16785	16431	16080	15732	15387	15044	14705	14367	14033	13701
19	17495	17135	16779	16425	16074	15726	15381	15039	14699	14362	14027	13695
20	17489	17129	16773	16419	16068	15721	15375	15033	14693	14356	14022	13690
21	17483	17123	16767	16413	16063	15715	15370	15027	14688	14351	14016	13684
22	17477	17117	16761	16407	16057	15709	15364	15022	14682	14345	14011	13679
23	17471	17111	16755	16402	16051	15703	15358	15016	14676	14339	14005	13673
24	17465	17105	16749	16396	16045	15697	15353	15010	14671	14334	14000	13668
25	17459	17099	16743	16390	16039	15692	15347	15005	14665	14328	13994	13662
26	17453	17093	16737	16384	16034	15686	15341	14999	14659	14323	13988	13657
27	17447	17087	16731	16378	16028	15680	15335	14993	14654	14317	13983	13651
28	17441	17082	16725	16372	16022	15674	15330	14988	14648	14311	13977	13646
29	17435	17076	16720	16366	16016	15669	15324	14982	14643	14306	13972	13640
30	17429	17070	16714	16361	16010	15663	15318	14976	14637	14300	13966	13635
31	17423	17064	16708	16355	16005	15657	15312	14971	14631	14295	13961	13629
32	17417	17058	16702	16349	15999	15651	15307	14965	14626	14289	13955	13624
33	17411	17052	16696	16343	15993	15646	15301	14959	14620	14284	13950	13618
34	17405	17046	16690	16337	15987	15640	15295	14954	14614	14278	13944	13613
35	17399	17040	16684	16331	15981	15634	15290	14948	14609	14272	13938	13607
36	17393	17034	16678	16325	15975	15628	15284	14942	14603	14267	13933	13602
37	17387	17028	16672	16320	15970	15623	15278	14937	14598	14261	13927	13596
38	17381	17022	16666	16314	15964	15617	15272	14931	14592	14256	13922	13591
59	17375	17016	16660	16308	15958	15611	15267	14925	14586	14250	13916	13585
40	17369	17010	16655	16302	15952	15605	15261	14919	14581	14244	13911	13580
41	17363	17004	16649	16296	15946	15599	15255	14914	14575	14239	13905	13574
42	17357	16998	16643	16290	15941	15594	15250	14908	14569	14233	13900	13569
43	17351	16992	16637	16284	15935	15588	15244	14902	14564	14228	13894	13563
44	17345	16986	16631	16279	15929	15582	15238	14897	14558	14222	13889	13558
45	17339	16980	16625	16273	15923	15576	15232	14891	14553	14217	13883	13552
46	17333	16974	16619	16267	15917	15571	15227	14886	14547	14211	13878	13547
47	17327	16968	16613	16261	15912	15565	15221	14880	14541	14205	13872	13541
48	17321	16963	16607	16255	15906	15559	15215	14874	14536	14200	13866	13536
49	17315	16957	16602	16249	15900	15553	15210	14869	14530	14194	13861	13530
50	17309	16951	16596	16243	15894	15548	15204	14863	14524	14189	13855	13525
51	17303	16945	16590	16238	15888	15542	15198	14857	14519	14183	13850	13519
52	17297	16939	16584	16232	15883	15536	15192	14852	14513	14177	13844	13514
53	17291	16933	16578	16226	15877	15530	15187	14846	14508	14172	13839	13508
54	17285	16927	16572	16220	15871	15525	15181	14840	14502	14166	13833	13503
55	17279	16921	16566	16214	15865	15519	15175	14835	14496	14161	13828	13497
56	17273	16915	16560	16208	15859	15513	15170	14829	14491	14155	13822	13492
57	17267	16909	16554	16203	15854	15507	15164	14823	14485	14150	13817	13486
58	17261	16903	16549	16197	15848	15502	15158	14818	14480	14144	13811	13481
59	17255	16897	16543	16191	15842	15496	15153	14812	14474	14138	13806	13475
60	17249	16891	16537	16185	15836	15490	15147	14806	14468	14133	13800	13470

Better books make better astrologers.
Here are some of our other titles:

Al Biruni
The Book of Instructions in the Elements of the Art of Astrology,
   translated by R. Ramsay Wright

Derek Appleby
Horary Astrology: The Art of Astrological Divination

C.E.O. Carter
An Encyclopaedia of Psychological Astrology

Charubel & Sepharial
Degrees of the Zodiac Symbolized

H.L. Cornell, M.D.
Encyclopaedia of Medical Astrology

Nicholas Culpeper
Astrological Judgement of Diseases from the Decumbiture of the Sick,
   and, **Urinalia**

Dorotheus of Sidon
Carmen Astrologicum, *translated by David Pingree*

Nicholas deVore
Encyclopedia of Astrology

Firmicus Maternus
Ancient Astrology Theory & Practice: Matheseos Libri VIII,
   translated by Jean Rhys Bram

William Lilly
Christian Astrology, books 1 & 2
Christian Astrology, book 3

Claudius Ptolemy
Tetrabiblos, *translated by J.M. Ashmand*

Vivian Robson
Astrology and Sex
Electional Astrology
Fixed Stars & Constellations in Astrology

Richard Saunders
The Astrological Judgement and Practice of Physick

H.S. Green, Raphael & C.E.O. Carter
Mundane Astrology: *3 Books*

If not available from your local bookseller, order directly from:

The Astrology Center of America
207 Victory Lane
Bel Air, MD 21014

on the web at:
http://www.astroamerica.com

Lightning Source UK Ltd.
Milton Keynes UK
UKHW040318080119
335170UK00001B/208/P

9 781933 303178